"The portfolio of essays is excellent, ranging widely, incorporating many different voices, stretching it seems (thankfully) beyond academia into the world of environmental practice, emphasizing the experience and first-hand accounts of the authors. I'm impressed, too, at the emphasis on personalizing educational experience, the developmental orientation, and the openness to multiple forms of learning. The essays challenge the reader to encounter some of the existential challenges that are inevitable in contemplating environmental issues. The authors are highly qualified, very experienced, and surely in Stuart Hill's case, among the founders of academic environmental studies."

Mitchell Thomashow, Director of the Second Nature Presidential Fellows Program, former president of Unity College, and former Chair of Environmental Studies at Antioch University

SOCIAL ECOLOGY AND EDUCATION

Social Ecology and Education addresses "ecological understanding" as a transformative educational issue: a learning response to emerging insights into social-ecological relationships and the future of life on our planet.

In the face of the existential threats posed by climate change, loss of biodiversity, pandemics and the associated ecological and social challenges, there is a need to extend our responses beyond scientific inquiry and technological initiatives. This book seeks to move the dialogue towards a deeper and broader understanding of the complexities of the issues involved. To achieve this, the book discusses issues rarely addressed through programs in "Education for Sustainability" and "Environmental Education," such as student defined knowledge systems, deep engagement with the implications of indigenous understandings, climate change as symptomatic of broad epistemological problems, social disengagement and differentiated barriers to meaningful change. This work is enriched by its focus on the learning and the learning systems that have led to our current predicament.

This book seeks to initiate considerations of this kind, to invigorate education for sustainable, equitable, healthy and meaningful futures. As such, this book will be of great interest to undergraduate and postgraduate students in a range of education and environmental courses.

David Wright is a senior lecturer in Education at Western Sydney University, Australia.

Stuart B. Hill is Foundation Chair of Social Ecology at Western Sydney University, Australia.

SOCIAL ECOLOGY AND EDUCATION

Transforming Worldviews and Practices

Edited by David Wright and Stuart B. Hill

Routledge
Taylor & Francis Group
LONDON AND NEW YORK

First published 2021
by Routledge
2 Park Square, Milton Park, Abingdon, Oxon OX14 4RN

and by Routledge
52 Vanderbilt Avenue, New York, NY 10017

Routledge is an imprint of the Taylor & Francis Group, an informa business

British Library Cataloguing-in-Publication Data
A catalogue record for this book is available from the British Library

Library of Congress Cataloging-in-Publication Data
Names: Wright, David (David George), editor. | Hill, Stuart B., 1943- editor.
Title: Social ecology and education : transforming worldviews and practices / edited by David Wright and Stuart B. Hill.
Description: New York : Routledge, 2021. | Includes bibliographical references and index. |
Identifiers: LCCN 2020013569 (print) | LCCN 2020013570 (ebook) | ISBN 9780367471088 (Hardback) | ISBN 9780367471095 (Paperback) | ISBN 9781003033462 (eBook)
Subjects: LCSH: Social ecology. | Educational sociology | Environmental education.
Classification: LCC HM861 .S624 2021 (print) | LCC HM861 (ebook) | DDC 304.2--dc23
LC record available at https://lccn.loc.gov/2020013569
LC ebook record available at https://lccn.loc.gov/2020013570

ISBN: 978-0-367-47108-8 (hbk)
ISBN: 978-0-367-47109-5 (pbk)
ISBN: 978-1-003-03346-2 (ebk)

Typeset in Bembo
by Taylor & Francis Books

With thanks to the pioneering work of Murray Bookchin and colleagues at the Institute of Social Ecology, and to pioneers in systemic approaches to learning, especially Richard Bawden, Graham Bird, David Russell, John Cameron and Judy Pinn. And with thanks to Christy, Louis and Arlo.

CONTENTS

ILLUSTRATIONS

Figure

Tables

PREFACE

As we finalise this collection, the threat of bushfires across the Australian continent is receding. The fires started in mid-autumn 2019, first in Queensland, then on the ranges and coast of northern New South Wales (NSW), then to the north-west of Sydney across vast areas of national park, then the south-west of Sydney, down the NSW south coast and deep into the east coast of Victoria, and then up into and across the tablelands surrounding and containing the Snowy Mountains. These fires have blitzed the land, homesteads, townships and lives. Add Kangaroo Island, the Adelaide Hills, large areas to the south and west of Perth and beyond. Over 10 million hectares burnt, 3000 homes destroyed, 32 lives lost, billions of both trees and wildlife killed, landscapes ravaged, emotions savaged, communities broken and confidence in the future thrown into disarray over a three month period of fire trauma. In the process, insight into the long-predicted impact of climate change has been delivered to a nation that is woefully underprepared, and still politically in denial both about the nature of climate change and about the radical, deep-rooted cultural changes that need to be enabled. Not only has Australia's political leadership been unwilling to address these predictions, but in many notable instances the significance of the actual fires themselves has been downplayed by those in positions of authority. Where to from here?

Additionally, as we finalise this collection, our university (and, we suspect, many other universities) are struggling to build the structures that will allow a trans-disciplinary study of social-ecological systems to be offered to students. Ill-informed assumptions, territorial fears, individual ambitions and structural inflexibilities conspire to make an obvious need exceedingly difficult to cater for and be supported in the long term.

Social ecology, as it has been taught in Australia and elsewhere, is a broad, deep and tangential (out of the box) area of study. At its heart it imagines conversations across disciplines about social-ecological interrelationships. It sees humans, individually and collectively, as the problem. We are the agents of our consequence. It is our choices that have brought about the unfolding ecological crisis. We are it and it is us. Why then is it so difficult for us to acknowledge this and act in response? What hinders our actions? What compromises our relationship with these most demanding of issues? How are we going to learn to respect the knowledge systems that reveal threats to our survival and enable us to chart a more sustainable and equitable course into the future? Learning of this kind is our subject matter in social ecology and in this book. Inevitably this subject matter exceeds the boundaries of this collection, but we cannot ignore or fail to take responsibility for our learning and, to an extent, we do so here.

Considerations upon social-ecological interrelationships necessarily initiate deep questions about the consequences of our ways of thinking and acting. As a result, feedback systems become ways of knowing. Sensitivity to feedback requires not just more sensitivity but action. And then all of a sudden something can arise that tells us our sensitivity and our actions have been insufficient. Bushfires roar across thousands of hectares. Australian rainforests catch fire for the first time in recorded history. The summer holidays of millions of citizens of a prosperous and well-educated nation are disrupted by months of smoke haze and the direct impact of walls of flame. Tens of thousands are forced to flee unstoppable fires, thousands find themselves trapped on beaches along the eastern seaboard, chest-deep in breaking waves as the surrounding foreshore and headlands erupt in ferocious burning heat. And as these apocalyptic images go viral and the world responds, the Australian government, Boer-like, prevaricates. And we, who work in education, must again confront the difficulties that our institutions have in valuing, prioritising and enacting learning systems necessary to change an inadequate status quo. Recalibrate, recalibrate, recalibrate. The learning that delivered us these circumstances is not sufficient to get us out of this situation. Transformative change is required: deep adaptation and whole system redesign. This is the context of this book.

Social ecology has been taught at Western Sydney University (WSU), previously the University of Western Sydney, since 1987 (following a similar program in social communication started in 1982). One of its most powerful outcomes is its community. Within this community, depth of connection generates empathy: shared honesty in response to change. This community spreads far and wide. It resists commodification. It extends. It draws people to it and holds them, lovingly, outside the definition of any university program. The program is a portal, but not the only one. It asks questions. It accredits responses. Many of those who have contributed to this collection have responded to the questions generated by the feedback systems that social ecology feeds into. In 1996, Stuart B. Hill arrived from Agriculture and Environmental Science programs at McGill University to be the Foundation Chair and Professor of Social Ecology. Central to his contribution to the program has been recognition that self-insight and self-care are integral to insight into and care for others and the Earth. David Wright, who bought an interest in creativity and embodied learning to the study, and Brendon Stewart, with a background in design, visual arts and Zen practice, have been longstanding staff members. Rachael Jacobs' work in aesthetics education is complemented by her political experience. Roseanna Henare-Solomona, Dale Hunter (writing with Stephen Thorpe), Kate Harris, Christy Hartlage, Werner Sattmann-Frese and Cathy McGowan are all former social ecology students. Each are powerful community activists as well as academic thinkers and writers, and their learning complements their practice. Cathy is also a former independent member of the Australian House of Representatives. Like Christine Milne, who is a former member of the Australian Senate and past leader of The Australian Greens, Cathy's activism continues post formal politics. Jen Dollin is a coordinator of sustainability at WSU and a contributor to the local and global network of Regional Centres for Expertise in Education for Sustainable Development acknowledged by United Nations University (UNU-IAS). Roseanna Henare-Solomona and Jo Clancy write powerfully from an indigenous perspective. Subarna Sivapalan and Ganakumaran Subramaniam bring depth of insight into sustainability in indigenous communities in Asia. Bob Jickling and Sean Blenkinsop bring deep learning from "the wild," where human nature nourishes ecological understanding. Ann Dale and Hilary Leighton, in their own way, write from "the edge," where "unexpected growth is possible and new life flourishes." Isak Stoddard does similarly, drawing on his long and diverse academic work across the natural and social sciences. This is a meaningful collection, arising in confronting times. We welcome you to it and look forward to the conversations it evokes.

ACKNOWLEDGMENTS

The editors welcome and acknowledge the support of Professor Michele Simons, the Dean of the School of Education, Western Sydney University, Professor Katherine Holmes, Head of the Centre for Education Research, Western Sydney University and Dr. Geoffrey Marnell from Abelard Consulting Pty Ltd. We appreciate the assistance and advice of all relevant Routledge staff.

PART 1

Transforming learning

1

(EDGE)UCATION BY DESIGN

Ann Dale and Hilary Leighton

Given the latest Intergovernmental Panel on Climate Change report (IPCC, 2018a), it is clear we are entering an era of profound transformation and change – ecologically, and consequently, socially and economically. The report finds that limiting global warming to 1.5°C would require "rapid and far-reaching transitions in land, energy, industry, buildings, transport, and cities" (IPCC, 2018b). Whether or not we can make the critical transitions in time to a carbon neutral economy remains to be seen, and it is clear that major change in current human development paths is also necessary (Burch et al., 2014; Moore et al., 2018; Shaw et al., 2014).

Yet, transformative change is another one of our messy, wicked problems as our research into the nature of change in current development paths has revealed (Jost et al., 2019). Unpacking change is difficult, as it is deeply normative with particular time dimensions. What one person regards as transformative change another may consider merely incremental change or even stasis. As well, the sense of urgency one lends to the issue (e.g. climate pollution), determines the pace and scale of change that decision-makers deem necessary. In recent research concerning three of British Columbia natural resource conflicts, environmentalists viewed the issues as more pressing than many industry interviewees; whereas the former considered that time was of the essence, the latter felt the opposite (Clermont, 2018). Regardless of these differences, scientific evidence shows that transformative ecological change is inevitable (Rockstrom et al., 2009); however, it remains to be seen whether human institutions are capable of intentional interventions in current exploitist fossil fuel development paths to limit climate change to a 1.5°C threshold. Are there critical lessons to be learned from socio-ecological system dynamics that educators could use to open space for transformation to happen, what we are calling *learning at the edge* or, in other words, (edge)ucation?

Systems thinking teaches us that in the living world with its infinite variants, there are multiple equilibria with diverse tipping points, and change happens in ways that are difficult to predict. Transformation of systems – and we posit of peoples and cultures – is unique to each context. In the case of humans, much depends upon meanings made from lived experience and life trajectories, from knowledge and awareness. As (edge)ucators, we believe it is our obligation to bring a wide, diverse and divergent pedagogy and as Madeline Grumet said, "it is the work of the teacher to interrupt the familiar" (1995, p. 16). Our job is to guide students in our graduate certificate for sustainable community development to turn over unexamined beliefs and biases, check orientations and assumptions, and illuminate fears and hopes and dreams that shape us by how we perceive and emanate them. It is the work of the courageous teacher called to this task to "stir the pot" of the stew the students grew up in, to help them think about and pose good and difficult questions about their lives and about what is truly important. As pedagogues accompanying students, we must help them dig for that sense of aliveness and purpose, and to find their "yeses" no matter the discomfort they experience in the digging or what else they might find underneath the surface of things. This can help students access their own personal proclivities and resources, or what Dewey called their innate "powers" (1929/2009, p. 34) in order to cultivate and eventually integrate these larger energies into their lives, precisely the fuel they need to do the work ahead. Teachers who facilitate this type of inquiry face unknown knowledge, together with the students, and by observing this process, through what curriculum theorist William Pinar calls a necessary "pedagogy of listening" (personal communication, July 2014), are able to help shift and reshape the curriculum from an emphasis on the abstract of externalities as a kind of disembodied learning to a less codified, more subjective and embodied approach. To question business as usual, to disrupt old habits, storylines and thinking, to push the edges of the familiar, invites the necessary tipping-into-back-loops of collapse, darkness and gestation (as drastic as that may sound for both systems and humans) towards an eventual transformation (even maturation?), a reorganisation of thought and action.

Transformative learning is defined by Simsek (2012) as a process of deep, constructive and meaningful learning that goes beyond simple knowledge acquisition and that supports critical ways in which learners consciously make meaning of their lives. Mezirow (2009) argues it is becoming critically aware of tacit assumptions/expectations and assessing their relevance for making an interpretation. It also involves simultaneous learning on two levels – cognitive and affective. Bloom et al. (1971) defined the latter as including an ability to deeply listen, to respond in interactions with others, to demonstrate attitudes and values appropriate to particular situations, to demonstrate balance and consideration, and to display a commitment to principled practice on a day-to-day basis, alongside a willingness to revise judgment and change behaviour in the light of new evidence. Sipos et al. (2008) have another interesting understanding of transformative sustainability education, that is, where learning objectives are "organized by head, hands and heart – balancing cognitive, psychomotor and affective domains" (p. 68). Hart (2001) encourages "an education of inner significance" (p. 7), where transformational experiences are more likely to occur when a link is made, and capacity is built, between the interiority of the student and the external world. With its richly layered curricular focus concerned with a movement of depth over growth that looks deeply into subjects rather than at the surface of things (so often associated with the mediocrity of much of education), Hart suggests we move from a mere information exchange to open into the rich terrain of knowledge and intelligence. Carved from both "the dialectics of intuition and the analytic" (p. 2), and cultivated through meaning made from direct experiences, students can become more compassionate and understanding with the wisdom to act ethically and the passion to do so.

Therefore, a genuine approach to transformational education and learning focusing on sustainability, requires not only an equal embrace of both mythos (deep imagination) and logos (rational and critical thought), but an immersion into the environment itself involving encounters with the "Other," eliciting a sense of interrelatedness and compassion for other life. This can act as an antidote to the (still) prevailing "epistemological error" (Bateson, 1979) at the heart of the Western worldview, with its perception of separateness and rugged individualism. A deep identification with other life in this way may be viewed as an end in and of itself with processes and outcomes unique to each student and their individual context (s). Within a more formative and emancipatory educational context of being-in-the-world (Heidegger, 1927/1962), self-knowing often leads to self-actualisation, reaching out beyond the discourse that happens in the classroom to consider the human soul in conversation with the world. Self-actualisation (the ability to act in accordance to one's true nature in contribution and service) as we were seeing it, more often than not leads a person to change (for the better), and this is in and of itself, transformative (Sterling, 2008).

The phenomenology of going outside to walk edges and explore intersections frames the here-and-now experience yet at the same time as we go out, we tend to go in and reflect deeply into matters of the heart with rich and complex outcomes that open to life and our deeper potential too. Stretches of time set aside for contemplation and reflection that allow space enough to give voice to metacognition in its identification of perspectives, biases, and values, through the meandering nature of journal writing for instance, is vital and yet is often left out of traditional pedagogy. These practices tend to get buried in a kind of hidden or subversive curriculum that is either discounted as too personal or is relegated to spare time beyond "more important" class time (meaning they are a lot less likely to happen, if at all).

In our classes, we offer a way to counterbalance this with an activity affectionately called *Terra Incognito* where students depart from the familiar classroom to traverse the unknown territory of the city and find a place that compels them in some way so they may go there to sit and listen every day (in all kinds of weather, at different times). From this place, they write, reflect, write some more and often sketch in their journals.

Walking and writing the city in this way is an immersion of being-in-the-world through reflective, embodied practice. These "conversations" the students have with the world facilitate a depth and sense of place beyond what we could have ever possibly taught in any classroom. Given the complexity of the issues involved in sustainable community development and the need for transformative learning on multiple levels (including the personal), this practice has been important for integrating and synthesising learning where new connections could form and were found at the intersections between the inner and outer worlds, striking a balance between rational thinking and the deep imagination. We humans are open systems conditioned to relationship and, according to anthropologist and cyberneticist Bateson (1979), are designed as such to receive, interpret and respond, in patterns of repetition between incoming signals and flow through of information and energy and throughput/feedback in an endless spiral of give and take. This type of recursive walking-writing reflective practice stimulates a kind of soaking in of information, inviting new connections with an enhanced ability to identify the interrelatedness of the local and global, people and community, species and biomes, past and future, and often creates new, original, reorganised thought in the process. We encourage students to consider the widest range of expression for their individual journal pages – writing, map-making, drawing, poetry, rubbings, photographs, sketches, pasted-in found ephemera (etc.), as they study the deep meaning of the city in its full grit and glory. The curation of their submitted journal pages in many ways mirrors what was most psychologically critical to them in that moment. Illuminating what they were attracted to (or conversely repelled by) in the city, this practice carried them to the edges of their own curiosity, inklings of understanding, and sometimes toward profound meaning-making. For instance, a student drawn to a particularly large tree in the square at city hall had noticed not only the comings and goings of all the people under its wide and generous shade-giving branches, but how much this old maple provided in terms of stability to all ages and stages of life, how it had become a hub welcoming all at different times of day and night (never discriminating), always patiently holding that open invitation. It was not a mere coincidence that for this student – a woman who worked as a community convener herself, with embodied traits of generosity and patience – these understandings only became clearer to her once she had let herself be drawn to this majestic tree and had investigated what was actually happening there. Of course, she was seeing the world through her own lens of experience, and yet the tree mirrored for her precisely what she needed in the moment to feel more confidence in her pursuit of sustainable community development solutions for a kinder, more connected city.

In ecological systems, the most biodiverse areas are at the edges where ecosystems meet – for example, the brackish waters between ocean and river where there is always more life (and death) to be found in the mix. Such "edges" host the highest levels of biodiversity and transformation. When habitats blend, unexpected growth, possibility and new life flourish. Unlike the dense, slow-growing, mature forest for example, at the blurred and muddy fringe (under optimal conditions), so much happens and yet it isn't all pretty. Many die here in service to new life because of their fragility and lack of resilience in this unstable environment of flux and change. Indeed, such environments can be difficult, yet nothing is ever wasted. As old trees fall, they become a reordered nursery for the young to feed and steady themselves upon, creating something generative from the fecundity of decay. Mostly, this is a place teeming with life (and death), where so much happens all the time and change is therefore guaranteed. To the onlooker this can sometimes appear to be nothing but destruction, or that nothing much is happening at all. However, in terms of extraordinary growth, so much occurs slowly – just underneath the surface.

To extend this metaphor, to teach from the edges, and in edgy ways, means being willing to let some plans and ideas die (like overly prescribed outcomes) so that others can live (like unpredictable and heuristic learning journeys) and not least of all, allow enough time for things to be in flux and change below the surface. Sewall writes, "The edge between land and sea, like other ecological edge effects is teeming with life, with abundance, as species stretch between ecological zones, as if the world is more, always more" (1999, pp. 135–136). Although it can be difficult, standing at the edge is likely to be where the greatest freedom and wisdom can occur if we are willing to meet those edges with courage, humility and compassion instead of fear (Halifax, 2018). Similarly, in life, during times of transition – birth, death, marriage and divorce – we are offered some of the richest opportunities for meaningful change and initiation to occur, if we are willing.

Holling's (1986) model of ecosystem functioning – from growth to conservation, to release and reorganisation, later known as panarchy theory (See Figure 1.1) – in many ways can serve as an important metaphor for transformative classroom learning. Many human institutions are "stuck" between growth and conservation, sometimes moving to a little release of the system or to almost no fundamental reorganisation (Dale, 2001). One can transpose government transitions on this model, and similarly, the stages of grief. One can become stuck, using Kubler-Ross' framework (2005) of the stages of grief – denial, anger, bargaining, depression – avoid the hard work of acceptance to transformation a reorganisation of the new reality, a new identity (e.g. no longer being just someone's parent, husband or wife).

To apply this model to human development, the release phase and threshold for tipping, while often frightening, is necessary for the old identity and its familiar supports to be stripped from its reference points and anchors of safety in a kind of severance or breaking away from the steady state. At this point, we break down and collapse, fall into the back-loop experience of the "dark night of the soul" (Jung, 1945). In this necessary darkness, in a time of in-between, there seems to be nothing happening; however, this disappearance phase is where we liquefy, just as the pupa does before its transformation into butterfly. It is our naturally resilient natures that allow us to become more fluid and permeable in this phase and emerge through to renewal. Resiliency is also determined by our ability to absorb the shock of falling (and falling to pieces) if we pool from a wide range of novel responses and, in effect, reorganise through self-organisation and self-actualisation.

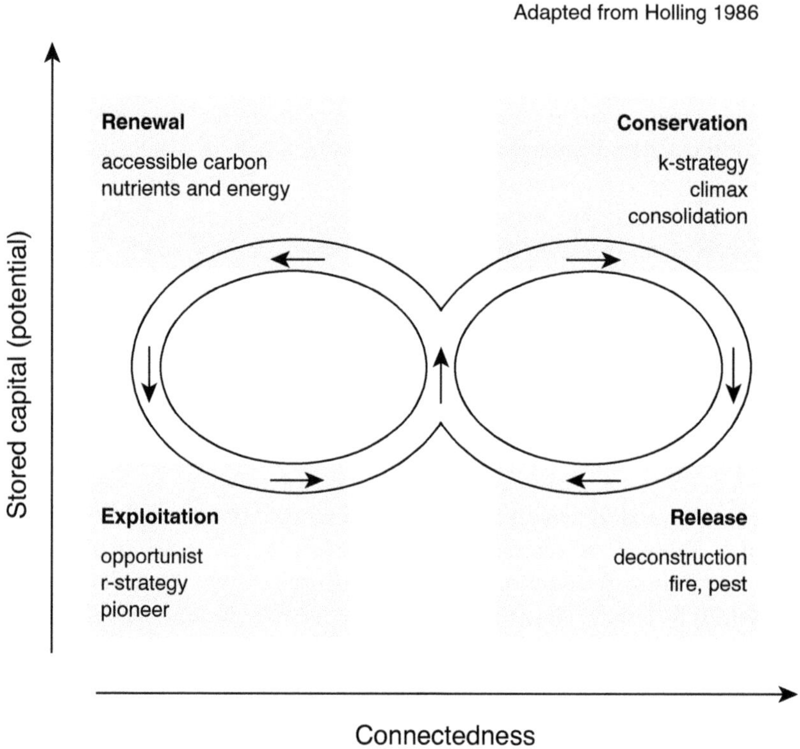

FIGURE 1.1 Holling's 1986 ecosystem dynamics model

In this liminal phase, capacity is built by localising, by making powerful new choices and finding strength in allies and community as well as self. Just as the butterfly reorganises into new communities of cells (called the "imaginal" cells) to become its flighted adult self. If we build capacity towards a vision for a compelling future by being willing to die to old ways that no longer serve, to learn from what has happened, then an initiation to a new level of integration, adaptation and identity can occur. Inevitably, in this renewal phase, we arrive at a clearer understanding of our purpose and of life's meaning, with access to the energy we need to act accordingly.

Besides the idea of creating a co-mutual learning environment, there are several other principles we practice. The first is that we want our students to awaken to how their own worldviews powerfully influence their thinking (their paradigms, myths and metaphors) and show them (let them experience) how to see the world through the lens of multiple perspectives in order to foster greater compassion and understanding of all "Others." We illustrate this using Jung's (1959b) idea of the "iceberg," where 95% of what we don't know is submerged, to link to the idea of the unconscious, a vast and richly laden realm of information and trapped energy, or what Jung (1959a) referred to as the "givens," the collective unconscious connecting the soul to something greater than any one of us can know. Ample time spent in nature immersion allows students to move past the notion of their own individualism and to connect to the vaster depths of community of intelligence – *Anima Mundi* or world soul (Hillman, 1989) – where perspectives can widen and shift and compassion deepens for all life beyond what any human teacher might try to convey.

Because there are many ways to enable learning – visual, auditory, kinaesthetic, sensory, and written – the educational experience should be inclusive of diverse styles and provide enough time for reflection and reflexivity. This includes finding creative ways of assessing what the students have learned and may include art, oral and written presentations, projects, etc. In addition to helping students construct their knowledge and intelligence from their experiences in the world, we aim to remain faithful to the etymology of education by helping them *educe* or draw forth and trust their wisdom from within, thereby transgressing the hegemony of one-way learning with its download of facts and data.

Integrating our two very different professional approaches (with the first author's initial experience as a civil servant, and the second author's expertise as an experiential educator, eco-psychologist and practicing psychotherapist), we do not regard our knowledge and expertise to be superior; rather, we use it to facilitate an integral space for co-learning between ourselves and with our students, registering important knowledge that continues to evolve from the inside and out in an effort to make the invisible more visible. This is congruent with the co-evolution of natural and human systems (Norgaard, 1994) within the Anthropocene era (Rockstrom et al., 2009). Such co-evolving, living, dynamic system interactions means that there is no single right answer, and not just one future, but multiple possible futures, with emergent phenomena occurring both slowly and rapidly.

If we return to Holling's ecosystem model, what are the lessons we can apply in the classroom? Just as with ecosystems, there are critical place, scale, limits and diversity functions to be respected (Dale et al., 2008; Dale, 2001; Dale & Newman, 2010; Newman & Dale, 2005). First, a safe space has to be co-developed where every student has the space to be themselves and feel as if they can be vulnerable to change – personally and through their learning. The second lesson is that place matters. Just as ecosystems are nested in and are a part of larger systems, so are human systems – we are nested within our homes, communities and institutions, and are subject to the limits of the planetary system. In practical terms, in order to best explore issues and barriers, dialogue and consult, convene public forums and presentations, and prototype and experiment, we need to create the conditions for students to systemically examine the issues not only from a place of the personal, but also from social, economic and ecological imperatives (Dale, 2001). Therefore, we have designed curricula to embrace the self-organising principles of living systems thinking where a system is recognised as not reducible to its components, as self-stabilising despite continual perturbations of in and through flowing information (and matter-energy), as having the innate ability to evolve in complexity amidst this flux, and as a whole in its own right as well as part of a larger whole or "holon" (Laszlo, 1996). Beyond anthropocentrism, this holistic approach brings the interrelatedness of all life and our relatedness to all things into focus and, in essence, suggests that what we do to the one, we, in effect, do to the other (Macy & Brown, 1998), placing greater responsibility upon students for their actions and inactions.

For us, this meant creating the necessary space and time in classes for students to critically examine and perceive the issues from various lenses and perspectives, and through multiple modalities in an attempt to reveal more of the interdependencies and correspondences that exist within the larger contexts and stories, rather than focus on what may be erroneously identified as "the problem." This also meant that we designed assignments that take our students outside of their own comfort zones, their familiar habits and habitual thinking, and physically out of the classroom to experience the world in situ.

Key cognitive domains that we focus on are applied knowledge: moving from the practicalities of sustainable community development to connection, regeneration, integrated planning, adaptation, reconciliation, and diversity, from local to global processes. To enable the students to understand and take meaningful action in relation to the challenges modern society is facing, we want them to have access to myriad lenses for applying theories, principles and practices, from architecture, art and poetry, science, nature, ecopsychology, anthropology, and dialogic inquiry. Curriculum content and design includes:

- upcycling (designing for cradle to cradle through circular economies; see McDonough & Braungart, 2013);
- biomimicry (borrowing genius from nature's designs to innovatively solve human problems; see Benyus, 1997; Harman, 2014);
- design thinking using a "maker space" (taking problems into fluid, constructive spaces of innovation, curiosity, risk-taking and empowerment combining empathy, creativity and rationality; see Doorley & Witthoft, 2012; Maguire, 2001; Manzini, 2015);
- bricolage (combining on-hand materials to craft the new; see Kincheloe & Berry, 2004; Wiseman, 2007), and
- open space technologies (allowing time and space for spontaneous, self-organising conversations around important issues; see Owen, 2008).

> No single element in the world is not bonded to, flying away from or catalytic with another element in the world…every ecosystem is an astonishing meeting, this conversation between various dynamics that contribute to the central conversation of life
>
> *(Whyte, 2011)*

A primary motive of "relatedness" implies that competencies developed by a systems approach that stresses an indissoluble unity of people and place in "conversations" with one another is necessary for us to handle the pressing challenges facing modern societies. "More often than not (in education), emotional [and] intuitive understandings are marginalized, considered too personal and unruly, or are discounted entirely" (Leighton, 2014, p. 139). To mitigate this, we firmly root more embodied and embedded practices into our curriculum design so that the living world and learning has a chance to meet and have far-reaching and catalytic conversations that facilitate emergence. Besides our co-teaching model, we use a Socratic style of sharing knowledge. This flips the classroom so that the wisdom of the students is on equal footing with their professors and encourages deeper conversations.

"What scholars now say – and what good teachers have always known – is that real learning does not happen until students are brought into relationship with the teacher, with each other and with the subject" (Palmer, 1998, p. xvi), and we would like to add – "and with – themselves." Our experience convinces us that personal change must come from the inside out rather than exclusively from external sources and expert knowledge. For this to occur, a whole person perspective of teaching must include the sensuous body, the reasoning mind, the emotional heart, and the ineffable, imaginative soul, with attendance to the student's experience and interests through divergent practices that build on their "capacity for personal strength, self-knowledge, integrity, compassion and cooperation" (Todesco, 2012, p. 115). Effective and responsive teaching requires that we draw from the entire epistemological spectrum of cognitive, aesthetic, emotional, physical and spiritual intelligences to align meaning and purpose for what is essential to each student (Hart, 2001; O'Sullivan, 2008; Palmer & Zajonc, 2011; Selby, 2002; Sterling, 2001), while being "careful to provide a balance of creative, practical scientific competencies and artistic skills as well as environmental awareness toward the efficacy of that student" (Leighton, 2014, p. 312) in service of achieving an ecologically sustainable, equitable and spatially just, emerging new world.

Teaching is still a great undeveloped and vast region that is continually evolving to meet the world with its challenges and crises, as characterised by high levels of complexity and uncertainty. But from the edge of where we stand, it appears to be evolving much too slowly to meet the current needs. Our responsibility runs deep to help equip our students with ecological competence, compassionate understanding and hope, with stamina enough to build communities that can ethically challenge the status quo and offer a more resilient and respectful relationship with the world (Orr, 2004).

If we can move beyond a mechanistic reality and clear a space for the more mythic and liminal realms to be included in the discourse through art and reflection, making and experiencing, then we believe it is possible to ontologise curricula to meet those necessary edges for change and change-making.

(Edge)ucation by design is where emergent learning (grounded in place, limits, scale and diversity) borrows from the same conditions that exist within the structure and healthy functioning of ecosystems. It demands equal attention to the physical, mental, emotional, imaginative and cognitive environment of the classroom, in order to honour and educate the whole person. A pedagogy that seeks inquiry that is integrative of both the natural and social sciences, breaks down the silos between what is "hard" and "soft." It creates a safe place to ask the difficult questions, to work through differences, respect different perspectives, seek ways to bridge those differences, to meet, and mix, and transform. And, not least of all, it intends to empower students to move from knowledge to action when they leave the sustainable community development classroom, so that "rapid and far-reaching transitions in land, energy, industry, buildings, transport, and cities" (IPCC, 2018a) can take place. Educating for a complete person means that this person will find meaningful employment and influence institutions, because the world needs people who are courageous, adaptive and open to radical and ongoing change, people who are not strangers to walking at the edges where new life happens. It needs people who employ fluid faculties of thought, feeling and imagination by using rigorous critical thought and profound creativity in an effort to contribute to society, as true "citizens of the world" (Nussbaum, 2010, p. 7). Our collective and sustainable future depends on this.

References

Bateson, G. (1979). *Mind and nature: A necessary unity*. E. P. Dutton.

Benyus, J. (1997). *Biomimcry: Innovation inspired by nature*. HarperCollins.

Bloom, B., Hastings, J., & Madaus, G. (1971). *Handbook on formative and summative evaluation of student learning*. McGraw-Hill.

Burch, S., Shaw, A., Dale, A., & Robinson, J. (2014). Triggering transformative change: A development path approach to climate change response in communities. *Climate Policy*, 14 (4), 467–487.

Clermont, H. (2018). *The underbelly of economy versus environmental conflicts. Detangling sources of tension in contentious natural resource decisions*. [Doctoral dissertation, Royal Roads University]. VIURRSpace.

Dale, A. (2001). *At the edge: Sustainable development in the 21st century*. UBC Press.

Dale, A., Ling, C., & Newman, L. (2008). Does place matter? Sustainable community development in three Canadian communities. *Ethics, Place & Environment*, 11(830), 267–281.

Dale, A., & Newman, L. (2010). All things counter, original, spare, strange: Why are we so bad at difference. *Canadian Journal of Humanities and Social Sciences*, 1(1), 37–43.

Dewey, J. (2009). My pedagogic creed. In D. J. Flinders & S. J. Thornton (Eds.), *The curriculum studies reader* (3rd ed.), (pp. 34–41). Routledge Falmer. (Original work published in 1929).

Doorley, S., & Witthoft, S. (2012). *Make space*. John Wiley & Sons.

Grumet, M. (1995). The curriculum: What are the basics and are we teaching them? In J. L. Kincheloe, & S. R. Steinberg (Eds.), *Thirteen questions: Reframing education's conversation* (2nd ed.) (pp. 15–21). Peter Lang.

Halifax, J. (2018). *Standing at the edge: Finding freedom where fear and courage meet*. Flatiron Books.

Harman, J. (2014). *The shark's paintbrush: Biomimicry and how nature is inspiring innovation*. White Cloud Press.

Hart, T. (2001). *From information to transformation: Education for the evolution of the consciousness*. Peter Lang.

Heidegger, M. (1927/1962). *Being and time*. Harper and Row.

Hillman, J. (1989). *A blue fire*. Perennial.

Holling, C. S. (1986). The resilience of terrestrial ecosystems: Local surprise and global change. In W. C. Clark & R. E. Munn (Eds.), *Sustainable development of the biosphere*, Cambridge University Press.

Intergovernmental Panel on Climate Change Secretariat. (2018a). *Summary for policymakers of IPCC Special Report on Global Warming of 1.5° C approved by governments*. [Press release].

Intergovernmental Panel on Climate Change. (2018b). *Global warming of 1.5°C: An IPCC special report on the impacts of global warming of 1.5°C above pre-industrial levels and related global greenhouse gas emission pathways, in the context of strengthening the global response to the threat of climate change, sustainable development, and efforts to eradicate poverty*. IPCC and United Nations Environmental Programme.

Jost, F., Dale, A., & Schwebel, S. (2019). How positive is "change" in climate change? A sentiment analysis. *Environmental Science & Policy*, 96, 27–36.

Jung, C. G. (1945). The philosophical tree. In G. Adler & R. F. C. Hull (Eds., Trans.). *Collected works of C. G. Jung, Volume 13: Alchemical studies*. Bollingen Series. Princeton University Press.

Jung, C. G. (1959a). *Collected works of C. G. Jung, Volume 9 (Part 1): The archetypes of the collective conscious*. (G. Adler & R. F. C. Hull, Eds., Trans.). Bollingen Series. Princeton University Press.

Jung, C. G. (1959b). *Collected works of C. G. Jung, Volume 9 (Part 2): Anion: Researches into the phenomenology of the self*. (G. Adler & R. F. C. Hull, Eds., Trans.) Bollingen Series. Princeton University Press.

Jung, C. G. (1965). *Memories, dreams, reflections*. (A. Jaffe, Ed.). Vintage Books.

Kincheloe, J., & Berry, K. (2004). *Rigour and complexity in educational research. Conceptualizing the bricolage*. McGraw Hill.

Kubler-Ross, E. (2005). *On grief and grieving: Finding the meaning of grief through the five stages of loss*. Simon & Schuster.

Laszlo, E. (1996). *A systems view of the world: A holistic vision for our time*. Advances in Systems Theory, Complexity and the Human Sciences (2nd ed.). Hampton Press.

Leighton, H. (2014). *Wild (re)turns: Tracking the epistemological and ecological implications of learning as an initiatory journey toward true vocation and soul*. [Doctoral dissertation, University of Victoria]. UVicSpace.

Macy, J., & Brown, M. (1998). *Coming back to life: Practices to reconnect our lives, our world*. New Society.

Maguire, M. (2001). Methods to support human-centred design. *International Journal of Human-Computer Studies*, 55(4), 587–634.

Manzini, E. (2015). *Design, when everybody designs: An introduction to design for social innovation*. MIT Press.

Maslow, A. (1968). *Toward a psychology of being*. D. Van Nostrand Company.

McDonough, W., & Braungart, M. (2013). *The upcycle: Beyond sustainability—Designing for abundance*. North Point Press.

Mezirow, J. (2009). *Learning as transformation: Critical perspectives on a theory in progress*. Jossey-Bass.

Moore, A., King, L., Dale, A., & Newell, R. (2018). Toward an integrative framework for local development path analysis. *Ecology & Society*, 23(2). https://doi.org/10.5751/ES-10029-230213

Newman, L., & Dale, A. (2005). Large footprints in a small world: Toward a macroeconomics of scale. *Sustainability: Science Practice, & Policy*, 5(1), 9–19.

Norgaard, R. (1994). *Development betrayed: The end of progress and a coevolutionary revisioning of the future*. Routledge.

Nussbaum, M. (2010). *Not for profit: Why democracy needs the humanities*. Princeton University Press.
O'Sullivan, E. (2008). Finding our way in the great work. *Journal of Transformative Education*, 6(1), 27–36.
Orr, D. (2004). *Earth in mind: On education, environment and the human prospect*. Island Press.
Owen, H. (2008). *Open space technology. A user's guide* (3rd ed.). Berrett Koehler Publishers Inc.
Palmer, P. (1998). *The courage to teach: Exploring the inner landscape of a teacher's life*. Jossey-Bass.
Palmer, P., & Zayonc, A. (2011). *The heart of higher education: A call to renewal, transforming the academy through collegial conversations*. Jossey-Bass.
Rockstrom, J., Steffen, K., Noone, A., Persson, A., Persson, S., Chapin*et al.* (2009). Planetary boundaries: Exploring the safe operating space for humanity. *Ecology and Society*, 14 (2). http://www.ecologyandsociety.org/vol14/iss2/art32/
Selby, D. (2002). The signature of the whole: Radical interconnectedness and its implications for global and environmental education. In E. O'Sullivan, A. Morell, & M. A. O'Connor. (Eds.), *Expanding the boundaries of transformative learning*. (pp.77–93). Palgrave.
Sewall, L. (1999). *Sight and sensibility: The ecopsychology of perception*. Jeremy P. Tarcher.
Shaw, A., Burch, S., Kristensen, F., Robinson, J., & Dale, A. (2014). Accelerating the sustainability transition: Exploring synergies between adaptation and mitigation in British Columbian communities. *Global Environmental Change*, 25, 41–51.
Simsek A. (2012) Transformational learning. In N. M. Seel (Ed.), *Encyclopedia of the sciences of learning*. (pp. 3341–3344). Springer.
Sipos, Y., Battisti, B., & Grimm, K. (2008). Achieving transformative sustainability learning: Engaging head, hands and heart. *International Journal of Sustainability in Higher Education*, 9 (1), 68–86.
Sterling, S. (2001). *Sustainable education: Re-visioning learning and change*. Green Books.
Sterling, S. (2008). Sustainable education: Towards a deep learning response to unsustainability. *Policy & Practice. A Development Education Review*, 6, 63–68.
Todesco, T. (2012). *Integrative and transformative learning practices: Engaging the whole person in educating of sustainability*. [Master's thesis, University of Victoria]. UVicSpace.
Whyte, D. (2011). *Life at the frontier: The conversational nature of reality*. [Video]. Youtube. https://www.youtube.com/watch?v=5Ss1HuA1hIk
Wiseman, B. (2007). *Levis-Strauss, anthropology and aesthetics*. Cambridge University Press.

2

TEACHING SOCIAL ECOLOGY

David Wright

Introduction

To introduce this discussion, I would like to tell a story. It is about my initial meeting with social ecology as a university-based study domain.

> Studies in the Master of Applied Science (Systems Agriculture) and the Master of Applied Science (Social Ecology) were taught, in 1995, through an intensive, residential model. Students would come from their home location and stay for four days on the campus of the University of Western Sydney: Hawkesbury to be launched into the Learning Projects that formed the basis of the programme. About 120 students were present for my first intensive as a newly appointed staff member. The first group session was an informal one where staff sought to encourage student interaction as a first step to the collective inquiry process. A staff member from Systems Agriculture mounted the stage and addressed the students. He spoke informally, welcoming everyone and celebrating their arrival. His aim was to share insight into who was there and to put people at ease, interacting and communicating, laughing if possible. He set up an exercise to move people around the hall: "Those who live locally to the left, those who have travelled from a distance, to the right." "Those who have children at home to the left, those without, to the right." Then, "those doing Social Ecology, to the left, those doing Systems Agriculture, to the right." The numbers were roughly equal. Then, "those who know what social ecology is, to the left and those who don't, to the right." About 115 people moved to the right side of the hall. A tentative five moved to the left. Laughter prevailed. I also moved, tentatively, to the right. Hmmm, I thought. Hmmm.

Looking back, I assume my appointment to teach in the Social Ecology programme was a consequence of a perception that I could contribute to the *social ecology* of the teaching and learning group as much as the *Social Ecology* we taught. This, I came to understand, was central to social ecology. I was not an expert, with a commanding overview. I was, potentially, an active participant in, and contributor to, an unfolding teaching and learning process. I can only assume my background in creative practice, systems theory and formal teaching and learning contributed to this, that it was seen as preparing me for a process taught, at postgraduate coursework level, through five key modules named "Learning," "Designing," "Communicating," "Researching" and "Ways of thinking." Interestingly, "Social Ecology" was not named subject matter in this programme, at that time. The term was used to focus attention on the relationships between modules: to acknowledge the complex relationship between people and the unfolding complexities of the social, emotional, intellectual and physical environments we inhabit.

Social ecology was then described by staff member David Russell as "a way of integrating the practice of science, the use of technology, and the expression of human values … [in] pursuit of designing activities that result in self-respecting, sensitive and social behaviours, which show an awareness of social and ecological responsibilities" (Russell, 1994, p. 148). There is a clear focus here on human meaning-making systems. In this regard social ecology was, in 1994, a theoretical construct, but not taught as such. Instead, teaching in the postgraduate coursework programme was designed to draw attention to social-ecological relationships within lived experience via the key modules. And the intention in the construction of student projects (for the programme was, and still is, taught through a project-based model of learning), was to engage students in the application of this way of thinking to real world encounters. Thinking was presumed to precede action, and perspectives upon personal experience within relationships were seen as pivotal. How are these experiences understood, represented and communicated? How are they enacted? What is the relationship between this and "learning"? Accordingly, an "action plan" or "project proposal" was central to each student's progress. This, the 1995 Master of Applied Science (Social Ecology) Student Handbook advises, "will emerge naturally as you review your learning, focus on the competencies you now need to develop (further) and decide how you wish to proceed." It requires a "visioning process, a further developing of your awareness of the ever-changing interrelationships between you, your work context and the people you work with" (Social Ecology Staff, p. 14). This was followed by the project itself. Thus, from its earliest days social ecology was not taught as subject matter; rather processes were designed to facilitate the emergence of an applied awareness of the social-ecological.[1]

Now, more than 20 years on, I and my Social Ecology colleagues continue to grapple with the challenge of facilitating the emergence of a social-ecological consciousness. This continues to be based on an assumption that such learning cannot be taught, but that skilled educators can assist it to arise through guided reflection upon personal experience: experiential learning, grounded in critical contextual inquiry (Freire, 1993; Kolb, 2005; Mezirow & Associates, 2000; O'Sullivan, 1999).

Hill (2011; updated in his chapter in this volume) describes Social Ecology as "the study and practice of personal, social and ecological sustainability and 'progressive' change based on the critical application and integration of ecological, humanistic, relational, community and 'spiritual' (the unknown) values to enable sustained wellbeing of all" (p. 19). He describes this as a provisional definition only. He continues:

> by stating my latest provisional thinking on the values that I consider central to social ecology I am hoping to encourage others to do likewise, partly to help me to further develop my own understanding. I am making the following statements not to say that this is how it is or must be, but rather that this is how it seems to me at this moment in time. It is my current story, my collection of narratives that make some sense of my experiences as a social ecologist. Working with such embodied stories is also central to my practice as a social ecologist.
>
> *(p. 20)*

In 2013 staff working in the programme agreed to introduce and teach a unit specifically named "Social Ecology" in the postgraduate coursework programme then on offer, the Master of Education (Social Ecology). It was argued that a unit such as this – building on the deliberately tentative depiction offered by Hill – could assist students by providing a structural and procedural framework for the programme: a base that would contribute to coherence within and between the subject-based units that would follow in the course on offer.

The teaching

It was thought that early in the teaching of the unit that the notion of Social Ecology as a discipline would need to be problematised. There is a structural logic to this: any mode of thinking needs first to question the basis of its construction, as well as constructions placed upon it. It is also a response to current debate. Daniel Stokols (2018), who teaches Social Ecology at the University of California, Irvine, argues for "a more unified analytic framework that … brings together social ecology's diverse concerns encompassing the natural, built, sociocultural, and virtual spheres of environmental influence on people's interactions with their surroundings" (p. 11); in effect, a global discipline. Although global commonalities can be valued, the importance of the local also requires recognition. In this respect, it is arguable that any social ecology programme needs to be discussed in relation to its own particular social ecology: that the local requires responsibility. In his work, Chet Bowers (1999) urges focus on local communities, local histories and local environmental practices. This can be best known through, Bowers suggests, greater awareness of place-based culture, tradition and "elder knowledge." Likewise, Sobel (1996) argues that we need to understand what we have local access to: "we teach too abstractly, too early" (p. 5), he asserts. Similarly, Grunewald (2003) cites Wendell Berry: "Unless one is willing to be destructive on a very large scale, one cannot do something except locally, in a small place" (cited in Grunewald, 2003, p. 633–634). For these reasons, in our local version of Social Ecology we ask students to evaluate the argument that the social ecological circumstances of the offering of any programme in Social Ecology are necessarily interwoven into its specific content and process.[2] The purpose of this is threefold. As well as localising discourse, this approach offers an opportunity to depict social ecology as a way of understanding unfolding relationships in a changing world based on local experience, rather than a static body of knowledge determined through a global discourse. It invites students' active engagement in that which is taught: its assumptions, its conceptualisation, its influences, its processes and its validity. Additionally, it is a means of enrolling students in the delineation of the experience they have entered into. It aids their recognition that it is their social ecology, and that their participation contributes to its formation and the means whereby it is understood, represented, communicated and enacted.

Accordingly, in our teaching (the Western Sydney version) the history of Social Ecology and the various iterations enacted around the world are discussed because of their significance, but none are presented as definitive. Although the notion of a discipline is actively resisted, the depiction by Murray Bookchin of social ecology as "concerned with the most intimate relations between human beings and the organic world around" (Bookchin, 2002, para. 5) is celebrated. It is also interpreted contextually, along with that of other past and present contributors to the discourse (Hill, 2011; Wright & Hill, 2011).

Invariably, and necessarily, this leads to questions around "our social ecology." And stories are told, often apocryphal. These locate the programme historically and orient it thematically. One is of a researcher in agricultural extension who ventured into country towns to inquire into the viability of farming in rural and regional Australia. This led to surprising insights: it was not the prices earned for farm produce that farmers identified as principal concerns, nor the vagaries of weather. Crucial were social relationships, principally familial. It was significant that so often children had to leave home to progress through school: that there were no teenagers around. The unpredictability of farm incomes required women to go into town to become the mainstay of the family, as nurses, teachers, pharmacists, shop assistants and more. Suicide rates among male farmers are among the highest in the land. The stresses of bank financing break families apart. These are social-ecological relationships. They are not issues of agriculture or governance or responses to scientific research. Rather, they are personal encounters with, and actions in response to, the social and environmental determinants of experience. They are stories told, brows wiped, scars earned, hearts uplifted, hearts broken.

Recognition of, and responses to, this depth of encounter became central to the programme in Social Ecology that was initiated at the then Hawkesbury Agricultural College (later the University of Western Sydney, and now Western Sydney University). These insights were theorised and positioned discursively

(Bawden & Packham, 1998; Bawden, 2004). They were interrogated through reference to different communities and environments and stories of experience (Mulligan & Hill, 2001; Wright, Camden-Pratt, & Hill, 2011; Wright, 2013). And although it was placed within context, the personal dimension of the study was never diminished. Hill (2011) argues:

> Most … holistic frameworks for understanding and acting in the areas of sustainability and change use as their three main categories economics, society and the environment… the common neglect of the personal supports the widespread perception that our problems can only be solved by heroes (mythologised rather than real people), particularly politicians and scientists, rather than problem solving (and, more importantly, prevention) being a collaborative project that requires all of our contributions.
>
> *(p. 20)*

Neville (2011) argues similarly:

> If it is indeed true that we are witnessing in ourselves and our students an emerging consciousness that has the capacity to face the present planetary emergency without denial or defensive reaction, and if we know how to make use of the new possibilities for human connection across nations and across cultures, our local and apparently insignificant contributions may make a difference.
>
> *(p. 69)*

More than any theorist, the personal experience of relationship is introduced as the spine of the study. O'Sullivan (2011) describes this as an interactive process. He argues that greater engagement with the natural world enables humans to "enter a relationship with the natural world that honours the deep subjectivity and interiority of all aspects of reality" (p. 40). By interrogating the subjectivity through which phenomena are known there are opportunities to appreciate the contextual base of knowing. Identifying that contextual base as representative of a network of interrelationships leads to the insight suggested by Gregory Bateson's oft cited reference to "the pattern that connects" (Bateson, 1979, p. 8). This is the basis of his "ecological epistemology" (1972, 1979). O'Sullivan (1999) argues this is Bateson's contribution to a "reconstructive postmodern vision." Its recognition of the ecological determinants of knowledge incorporates an appreciation of the systemic boundaries of social-ecological experience: the materiality of our planet. By virtue of our participation – our interdependence – we are intimately related to the systems that sustain the relationships through which we know ourselves. This places Social Ecology staff in the position of active and conscious contributors to relationships that sustain knowing. Of necessity, they are required to do more than teach Social Ecology; they are required to consciously contribute to a social-ecological way of thinking, understanding and acting: to work through a relationship-based approach to learning. Fritjof Capra (1996) draws on Maturana and Varela (1992) to describe this as "bringing forth our world": the ongoing creation of perspective and experience through insightful contributions to the social-ecological relationships we inhabit. This arises in the context of the recent recognition that we are now living within the Anthropocene epoch,[3] the result of previously unimagined (and for some, still unacknowledged) changes in human–earth relationships.

Systemic origins

As stated, the origins of the Social Ecology programme at Western Sydney University lie in agriculture, most specifically in systems agriculture. Bawden (2004) describes systems agriculture as a response to the limitations of a techno-scientific approach to agriculture in Australia. He observes that, the "systems idea" was thought to draw attention to "the complex inter-connectedness of most of the impacts that were (then) being observed, independently of each other."

> It was seen as a study of "a bounded wholeness," comprised of embedded and mutually associated subunits, that displays properties that are emergent through the inter-actions of those "lower order wholes," as well as those between the "whole" and the "higher order unit of wholeness" in which it is, in turn, embedded.
>
> *(Bawden 2004, p. 58)*

As in systems agriculture, early attention in social ecology was placed on systems thinking and active, experiential learning processes. In effect, how learning of a systemic kind can be arrived at. Necessarily, questions of this sort gave rise to considerations upon the immersion of the subject in the object (and the object in the subject): the learning. Consequently, learning came to be seen as embodying self-observation of self-organisation. This form of radical constructivism (von Glasersfeld, 1996) contributed to the early emphasis upon self-organising systems theory, drawn from research into the biology of cognition (Maturana & Varela, 1992), and cross-disciplinary analyses of the systemic roots of consciousness (Capra & Luisi, 2014). Importantly, this occurred at a time when the humanities and social sciences were absorbed in substantive challenges to orthodoxies through postmodernism and post-structuralism. This work can be seen as incorporating, while extending, that critique through recognition that the constraints of the physical universe – the system underpinning human consciousness – require incorporation in issues of perspective, structure and practice. Thus, in our work, environmentalism took a social-ecological turn by incorporating a problematisation of how the environment is known and related to. These issues were not seen as static, but as constantly unfolding. The more we learn about our experience of our relationships with the physical universe the more we question our insights, practices and assumptions: in effect, our history and our thoughts about our future. The Anthropocene demands this. Accordingly, Wright and Hill (2011) argue that from its earliest days "personal understanding" was central to all Western Sydney courses in Social Ecology. This understanding was contextualised and "applied to locales, practices and fields of knowledge with which the learner was directly concerned" (p. 4).

> [This] emphasized the centrality of relationships, and the importance of considered reflection ... It encouraged learning through participation ... [and, significantly] early ... leadership was provided by women members of staff ... the focus on feminist epistemologies, experiential and process-based perspectives on learning... contributed to the moulding of a personalised approach to research training.
>
> *(p. 4)*

But how can this done, and how can a "social-ecological perspective" be drawn from an experiential learning process?

Personal systems

As argued, work in the biology of cognition has had significant impact in Social Ecology. And although this is a theoretical perspective, it is taught for its application: for the transformation in learning arising as a consequence of an education in perspective. Although the biology of cognition is being taught, what is taught more overtly are stories of related experiences: "my learning through the biology of cognition." Key is reference to Maturana and Varela's (1992) concepts of "autopoiesis," "structural coupling" and "structural determinism." These are not taught for their exactitude, but as points of reference in a storytelling process. They are used to reflect upon ways in which "I observe my own learning" in relation to the world beyond. This includes the self, other humans, non-human life, the physical environment I find myself within, and the idea systems that contribute to my understanding of the ways of knowing I encounter.

Storytelling is a powerful form of communication, and stories of relationship can be particularly prescient in an educational environment. In my classes I often talk with students about performance poet Aaron Williamson, who happens also to be deaf. In performing his poetry, Williamson seeks to communicate his experience of sound to a hearing audience. In telling these stories I draw attention to the power of metaphor. In considering Williamson's fascination with his deafness, his encounter with "sound" – something he will never know as those with clear hearing do – I ask: what is my version of Williamson's ear? For I too am attempting to write and speak about things I cannot (ever) fully know. Like he, I have absorbed a considerable amount of "learning" about my subject matter. Like he, I have moved beyond the first-order activity of "learning" about that which I can never fully know, for new knowledge is constantly emerging, to the second-order activity of "learning about my learning." It is only through learning about my learning that I can critically engage with the social-ecology of that (and any) subject matter.

It is through reference to Williamson that I attempt to articulate a powerful subjective pursuit. I can look at his work and see not my pursuit of something I cannot fully know, but a pursuit *like* my pursuit of something I cannot fully know. Because language communicates in relationship, creativity is used to access the language most appropriate to specific relationships. Here metaphor is a linguistic form used to communicate "what it is like" (in relationship) because *what it is* can only be known in the experience, by the person who has the experience (and that knowing arises before language is used to stabilise and share a representation of the experience). Through reference to Williamson I suggest to my students that learning becomes, as McNiff (1992) says, are a contextually driven "unending process of attunement" (p. 57). Hillman (1996) depicts this conflict well. He argues that it is becoming "more and more difficult to make a cut between the psyche and the world, subject and object, in here and out there. I can no longer be sure," he says, "whether the psyche is in me or I am in the psyche as I am in my dreams, as I am in the moods of the landscapes and the city streets" (p. 154).

Another storyteller who builds on the base provided by biology of cognition is Andreas Weber (2016). Weber was a student of Francisco Varela and a contemporary of David Abrams. He writes of what he calls the "biology of the feeling self" (2016, p. 2). He argues that the process of biology is:

> discovering subjectivity as a fundamental principle throughout nature … even the most simple living things – bacterial cells, fertilized eggs, nematodes in tidal flats …value everything they encounter according to its meaning for the further coherence of their embodied self.
>
> *(p. 3)*

He describes this as a "poetic ecology" because "it regards feeling and expression as necessary dimensions of the existential reality of living organisms." He argues that recognition of this kind constructs ethical responsibilities. "To be able to imagine this sort of ethics we must remember that the feeling body is the ground zero of any morality" (p. 335). Like Abrams, he seeks to make himself vulnerable to "the poise that comes from living in storied relation and reciprocity with the myriad things, the myriad *beings*, that perceptually surround us" (Abrams, 1996, p. 270).

This emphasis upon learning processes is therefore a direct consequence of the exploration of a systemic perspective. Not only does the course posit a systemic – social-ecological – perspective, but it asks students to test this perspective in their workplace, community, network and family: to apply it *in situ*. For this reason, I want to extend this discussion by looking at projects submitted by two postgraduate, course-work students for assessment in the unit Social Ecology. These students, I will call DA and SD.

Project-based learning

In her principal assignment for this unit, DA details a project she initiated in her work with Aboriginal girls in a school established for troubled students with patterns of suspension or expulsion because of truancy or violent behaviours. DA writes, "The most important thing I have discovered working with these girls is that relationship is everything." She cites arguments pointing to the need to identify and engage with positive attributes of students and observes that students "respond to how they are related to in the moment." With this in mind she adds, "I cannot expect to change the girls' behaviour ... All that I can do is take responsibility for my own words and actions." Accordingly, DA initiates a project involving greater experience of the outdoors. "I want to support them to grasp other ways of encountering the world and to assist them to have choice." She observes:

> I have found that the girls respond really well to being out and about in nature, they are happier, more relaxed, they smile and laugh more, there is a sense of connection and relief in their being. What I would also like to acknowledge is my own response. I feel more peaceful in nature, and there is a sense of space and less tension as there are fewer things to distract or cause angst. This in turn makes me feel more relaxed... the sensation feels more authentic and less strained. The students sense this mood change within me and respond.

DA tried to talk with the girls about the experience, but encountered reluctance:

> The girls do not really like to talk about anything that may expose them too much... Mostly the girls answered with a "No tit (sister) why you asking this stuff for?"... I dropped the interviews and decided on a more relaxed, informal... way of collecting data, this included incidental comments and observations of behaviours... I discovered that many of the girls do have special places they like to visit. On an occasion when we were driving one of the girls asked me to detour to take them past a special place, it was a small headland with a cliff. The girl showed the other girls the area and made comments like "I love coming here, my mother likes to drive out here," and "I love looking out at the sea it makes me feel calm."

The consequence is:

> We have increased our trips into the outdoors, the girls seem to crave it now, getting out and in nature as often as possible. We have taken to packing a picnic and heading somewhere by the river or the park most days. The girls instigate this and encourage us, the teachers... (they) put their phones away, talk, laugh and play, they ask questions and listen to the teachers. This has become like a circle time when we are all present and part of the group... it's like breathing out, even if only for a while.

This work is grounded through reference to key readings and summarised. Among the conclusions DA lists are "[the girls] are being challenged physically, and stimulated and calmed mentally by the natural world. They get to smile and laugh more, increasing the time they feel joyful." This is interesting, particularly in light of the motivation for the study that DA provides. This "comes from a place of wanting to deepen my commitment to my work place and marry my social ecology study in an attempt to bring my life's interests together and make my participation in the world more authentic."

DA's project typifies projects undertaken in this unit. It is based around an activity designed to enrich a work or community setting, and the experience and perspective of the researcher is central to learning that is arrived at.

Student SD did something similar. However, although the unit Social Ecology is designed to be undertaken at the start of a student's programme, circumstances led to SD doing the unit at the conclusion of her studies. This influenced her thinking considerably. She writes, "As I reach the end of my studies and prepare to enter a new stage, I wonder: am I a 'Social Ecologist' and, if so, what does this mean I *do* now? And how might I do it?" Initially, she perceives the work she proposes in the unit as an opportunity to "create heartfelt, personally sustainable … work as a way of participating in the world." This way is self-oriented. "It is in the realm of the personal that I have repeated the exhausting pattern of stumbling to a halt when trying to create this change in times past." Consequently, "thinking *surely I know enough by now*, I attempted to step into professional social ecologist mode." (Italics in original).

However, her orientation skewed inwards. "My undertaking did not go quite as planned. In fact, the more I planned in this linear, compartmentalised, rationalist way, the more uncomfortable I became and the further away I felt from enacting social ecology."

She says she "doubted my fit in what I can now see was shaping into an 'expert' role that focused on generating change 'out there'."

This led to a revised approach. Various considerations were launched before she "decided it was time to stop trying to calculate answers and begin again with questions." She asks:

> *So, what is it I long for, for myself and my loved ones near and far? What brings me energy? What resonates, personally and spiritually? Where do I feel I can move already? How am I when I am relating to others in a way that feels grounded in mutuality and interconnectedness?* (Italics in original).

Her response is anything but linear. "The questions struck home like lightning bolts. With this type of intuitive yet strategic questioning, I began to uncover," she says, quoting Peavey (2000), "deep desires of the heart, rather than communicating information already known" (p. 1). The process of relationship with self, she argues, "is inherent in my understanding and enactment of social ecology. When combined with powerful and strategic questioning … relationship takes on a rich conversational nature that has been a long time coming."

Guided by a question I asked in the early stages of the unit – *what do you need to learn and how are you going to learn it* – SD says that she needed to learn about "personal agency." Accordingly, she initiates what she calls an "Ecocreative action" a "poetic process of inquiry."

> Rather than building a path by planning content and mapping projects, I found a way to move forwards through creative action that crystallised values, traced desires and voiced a way of being and knowing in the world. This aligns with a social ecological approach as a way of "imagining, integrating and designing" (Wright & Hill, 2011, p. 5) my own participation.

Although the process that is initiated is interesting, involving as it does the development and distribution of a "planifesto" (which she describes as "a playful manifesto for the planet") and creative group work in natural settings, the reflections that form the conclusion of the projects go beyond the project itself. They indicate a considerable depth of learning. SD writes,

> On one level, this process has been directed towards planning future work. At a deeper level it has been strong work in the process of self-making… This path, rather than stepping into the role of The One Who Knows How, was about honouring vulnerability and resistance, deep listening, personal reflection and embracing creativity to generate a new way of being in the world… I have come to realise that during the years of study in social ecology, I've experienced something of an educational rehabilitation: I have restored or perhaps, more accurately, developed a sense of agency and a way of knowing that is inherently linked with both action and learning. I can't explain the depth of my gratitude for this. Tears have watered this path; not in sadness but in quiet, overwhelming grief and the healing of something parched… To say I'm a social ecologist is an invitation evoking a beautiful question from self and others; what is a social ecologist? I have heard this question many times. From this point on I welcome the conversation.

Despite teaching in this programme for many years, I have often been reluctant to call myself a social ecologist. Although I recognise and welcome the increasing use of the term "social ecology" to describe observable phenomena – the social ecology of education or the social ecology of national parks management, for example – I am ambivalent about claiming it as an identity. I have regularly argued that I see most value in the application of the term. That is, its use as an adjective – "a social-ecological" perspective, or way of thinking, or understanding – resonates more fully with me. It is not a stable thing; it is not a discipline or a container. It is an applied consideration: a way of knowingly participating in a set of relationships amidst change of a most profound kind. In this sense, my awareness of my own participation facilitates my awareness of my complicity in the epistemological formations that Social Ecology critiques. This construction, referred to widely as The Anthropocene, is not therefore something that is occurring out there, in the world. It is local, it is every day and it is personal. In effect, I am The Anthropocene. And I take responsibility. I am a social ecologist. There, I've said it. I am a social ecologist.

Notes

1 A unit "Introduction to Social Ecology" was taught by Professor Stuart Hill to students enrolled on the Bachelor of Applied Science (Social Ecology) program between 1996 and 2002. This unit, which built on an extensive literature base, was well received by students. Because it was taught to full-time undergraduate students the study could not be based around community experience, in the same way as the Masters' units.
2 The history of the naming of the UWS/WSU "Social Ecology" is discussed in more length in Wright and Hill (2011).
3 American biologist Eugene Stoermer and Dutch atmospheric chemist Paul J. Crutzen are generally recognised as being the originators, and popularisers, of the term Anthropocene, which acknowledges that we are now living in an era when human actions are having drastic effects on the Earth, especially its climate and biodiversity (Bonneuil & Fressoz, 2017)

References

Abrams, D. (1996). *The spell of the sensuous.* Pantheon Books.
Bateson, G. (1972). *Steps to an ecology of mind.* Ballantine Books.
Bateson, G. (1979). *Mind and nature.* Bantam Books.
Bawden, R. (2004). Angst, agoras, and academe: Reflections on an experience in conscious evolution. *The Journal of General Evolution,* 60(1–2), 53–66.
Bawden, R., & Packham, R. (1998). Systemic praxis in the education of the agricultural systems practitioner. *Systems Research and Behavioural Science,* 15, 403–412.
Bonneuil, C., & Fressoz, J-B. (2017). *The shock of the Anthropocene.* Verso.
Bookchin, M. (2002). Reflections: An overview of the roots of social ecology. *Harbinger: A Journal of Social Ecology,* 3(1), 6–11.
Bowers, C. A. (1999). *Changing the dominant cultural perspective in education.* In G. A. Smith & D. R. Williams (Eds.) *Ecological education in action* (pp. 161–178). SUNY Press.
Capra, F. (1996). *The web of life.* Harper Collins.
Capra, F., & Luisi, P. L. (2014). *The systems view of life.* Cambridge University Press.
Freire, P. (1993). *The pedagogy of the oppressed.* Continuum.
Grunewald, D. A. (2003). Foundations of place: A multidisciplinary framework for place-conscious education. *American Educational Research Journal,* 40(3), 619–654.
Hill, S. B. (2011). Social ecology: An Australian perspective. In D. Wright, C. E. Camden-Pratt, & S. B. Hill (Eds.), *Social ecology: Applying ecological understanding to our lives and our planet* (pp. 17–30). Hawthorn.
Hillman, J. (1996). *The soul's code.* Random House.
Kolb, D. A. (2005). *The Kolb learning style inventory – version 3.1.* Hay Resources Direct.
Maturana, H., & Varela, F. (1992). *The tree of knowledge.* Shambhala.
McNiff, S. (1992). *Arts as medicine: Creating a therapy of the imagination.* Piatkus Books.
Mezirow, J., & Associates. (2000). *Learning as transformation: Critical perspectives on a theory in progress.* Jossey Bass.
Mulligan, M., & Hill, S. (2001). *Ecological pioneers.* Cambridge University Press.

Neville, B. (2011). Chaos, reaction, transformation. In D. Wright, C. E. Camden-Pratt, & S. B. Hill (Eds.), *Social ecology: Applying ecological understanding to our lives and our planet* (pp. 64–70). Hawthorn.

O'Sullivan, E. (1999). *Transformative learning.* Zed Books.

O'Sullivan, E. (2011). Attempting an integral Earth story in the new century: Prospects for a twenty-first century education. In D. Wright, C. E. Camden-Pratt, & S. B. Hill (Eds.), *Social ecology: Applying ecological understanding to our lives and our planet* (pp. 31–40) Hawthorn.

Peavey, F. (2000). *Heart politics revisited.* Pluto Press.

Russell, D. (1994). Social ecology – Education and research. In L. Fell, D. Russell, & A. Stewart. (Eds.), *Seized by agreement, swamped by understanding* (pp. 147–165). Drs Fell, Russell & Assoc.

Sobel, D. (1996). *Beyond ecophobia: Reclaiming the heart in nature education.* The Orion Society.

Social Ecology Staff (1995). *Master of applied science (Social Ecology) student handbook.* University of Western Sydney.

Stokols, D. (2018). *Social ecology in the digital age.* Elsevier.

von Glasersfeld, E. (1996). *Radical constructivism: A way of knowing and learning.* Falmer Press.

Weber, A. (2016). *The biology of wonder.* New Society Publishers.

Wright, D., & Hill, S. B. (2011). Introduction: The emerging field of social ecology. In D. Wright, C. E. Camden-Pratt, & S. B. Hill (Eds.), *Social ecology: Applying ecological understanding to our lives and our planet.* Hawthorn.

Wright, D. (2013). Schooling ecologically: An inquiry into teachers' ecological understanding in 'alternative' schools. *Australian Journal of Environmental Education*, 29(2), 136–151.

Wright, D., Camden-Pratt, C. E., & Hill, S. B. (Eds.). (2011). *Social ecology: Applying ecological understanding to our lives and our planet.* Hawthorn.

3

TRANSFORMATIVE LEARNING PRIORITIES

Stuart B. Hill

Introduction

All aware, responsible thinkers and activists who want to enable meaningful change daily face the challenge of choosing what is best to say and do – in the face of widespread cultural and personal ignorance, misinformation, denial, postponement and even ridicule. And yet, paradoxically, underneath this surface expression, another self can often be detected, one that is yearning to "hear voices" that are able to clearly articulate what they remain confused about and that will enable their problems to be heard, appreciated, and possibly addressed.

Communicating with this *other self* is the challenge facing me as I write this chapter. I am now in my 70s and have spent a lifetime endeavouring to act responsibly and meaningfully (primarily within educational and agricultural systems), drawing particularly on my learning in ecology and psychology. While doing this I have endeavoured to keep in mind the complex internal and external enabling forces and barriers involved, some of which are relatively obvious, but most of which remain hidden and unknown. I am aware that this may be the last chance I have to publish my ideas relevant to this topic, so I will be sharing my thoughts with minimal censoring.[1]

Present challenges

- **Our species is in serious trouble**, and **most of our population appear to be largely unaware of how serious the situation is**. Our population and rates of consumption and waste production are continuing to grow, but our resource base is not. The readily accessible non-renewable resources have already been consumed, so it is getting harder and harder to extract those that are left and, as a consequence, there is declining return on investment. Renewable resources are being consumed faster than they can be replaced, and in ways that are destroying our soils, waterways and oceans, with rapidly growing losses of biodiversity and wild habitats. These ecosystems and their inhabitants are required for the maintenance of a planet in which our species can survive. Because awareness of the importance of this is yet to be widely acknowledged, by the time that the collapse of systems is recognised they may have already crossed thresholds where recovery is no longer possible (Bendell, 2018; Read & Alexander, 2019).[2] Community and personal wellbeing are also being eroded along with the losses of natural capital.

- **Few politicians are providing the leadership that is required**, and although voices in science, academia and business – based on reliable data – are arguing for significant changes in how we are living and managing resources, market-based decisions by our elected representatives continue to undermine ameliorative actions. Although most of the population is becoming concerned about this lack of leadership, they feel excluded from being able to change this situation.[3]
- This lack in both awareness of our situation and in political leadership **requires the development of bold educational initiatives** to enable our population to understand what is required for the sustainability of our species (and all other species), and to develop the competencies to make the necessary changes to our lifestyles and cultures, and particularly to our institutional structures and processes, including those concerned with education (Hill, 2012). The aim of this book is to address this need by advocating radical changes in education.
- Sadly, **our present educational systems are still largely preparing students for unsustainable futures**, stuck within neoliberal-dominated contexts (Greenwood, 2010; Fletcher, 2016; Henderson et al., 2018). Measures of success remain predominately growth in productivity, and individual profit and power, rather than ecological sustainability, equitable wellbeing of all life and meaningful caring relationships.

Barriers to progress, and some ways forward

The late Scottish psychologist R. D. Laing (1971) described a significant, yet rarely acknowledged, **psychological and systemic barrier to meaningful change**. Reflecting on his lifetime of work on enabling individual change, he concluded that *it is as if each of us has been hypnotised twice, firstly into accepting pseudo-reality as reality, and secondly into believing that we have not been hypnotised.* This personal (and cultural) adaptive behaviour is a coping response to discomfort and stress. Although it protects us over the short term from the responsibilities that flow from our experiences, over the longer term such adaptive responses contribute to our disempowerment, loss of awareness and our confusion about our visions and values – and to our being in denial about any of this. These losses of our potential give rise to a diverse range of subconsciously developed and actualised compensatory behaviours, such as attraction to symbols of power and the need to appear in control to mask our disempowerment. Such attractions are widely relied on in advertising, and are responsible for much of our growing consumerism and attraction to material and economic growth. Changing these behaviours would significantly contribute towards our living sustainably, and enabling such changes is central to the design and implementation of the transformative education that we are advocating in this book.

Within our societies, carrying out endless studies to document problems – another coping behaviour[4] – predictably rarely leads to significant meaningful change. Most actions, if they are eventually taken, tend to focus on back-end, shallow, reactive, short-term, single-factor, heavy-handed, de-contextual initiatives (like pesticides for pests and punishments for misbehaviour in schools). "Solutions" of this kind distract us from investigating the causes and the systems within which the problems have emerged.[5] Problems need to be recognised not as enemies to be eliminated, but rather as indicators of the mal-design and mismanagement of the systems involved (whether a farm or an educational institution). These need to be redesigned, taking into account the complexities involved. They should be better managed so as to enable wellbeing, meaningful learning, sustainability and the prevention and minimisation of problems – with an emphasis on informed and creative pro-action, rather than waiting to respond to problems when they are eventually recognized and acknowledged.

The other sobering challenge for me in writing this chapter is that **there is relatively little research literature that is broad and deep enough to inspire and guide** me in advocating educational systems in which we can all learn to live meaningful lives that are sustainable, equitable and nurture wellbeing. Despite many documents having hopeful titles, most primarily focus on critiques of the present situation, or on just a small part of the issue, with little or no content on what might be done to enable meaningful *whole system change*: from person to planet.[6] And most suggested changes tend to be fragmentary and adaptive. Often substitutes are sought for unsustainable, inequitable and harmful products and practices that can be used within existing systems when what is needed is radical whole system redesign and its ongoing testing and improvement against evolving agreed goals. This is where *servant leadership* (Greenleaf, 2002/1977; Spears & Lawrence, 2001) and *co-operacy* (Hunter et al., 1997) are required to enable clarification and consensus regarding goals, and the prioritisation of existing system redesign (and design of new systems) to achieve those goals.

Of particular significance for educators is the use of the arts and fiction (stories) to inspire and enable students to think critically and creatively about our history, present situations and possible futures. For example, concerning the challenges and *solutions* to climate change, whereas some students will respond best to the information provided in Naomi Klein's (2014) well-argued non-fiction book *This Changes Everything: Capitalism vs. the Climate*, [7] others may be equally inspired by Barbara Kingsolver's (2012) outstanding novel *Flight Behaviour*. For many years I used Daniel Quinn's (1992) creative novel *Ishmael* to enable deep transformation in my social ecology students' understanding of our relationships and responsibilities.[8]

Social ecology to the rescue

I believe that **social ecology** (the version that I championed at Western Sydney University)[9] **can provide a comprehensive framework for effective understanding and action**. This is because it includes the main interrelated areas that need to be considered when working with transformation and meaningful change: the personal, socio-cultural (including, but not privileging economic, political, technological and all other institutional structures and processes), and environmental and spiritual (for me, the vast unknown, and yet to be discovered and understood). My experience is that working to deeply understand our situation and act effectively (personally to globally, and immediately to trans-generationally) requires the support of such a holistic framework (Hill et al., 2004; Hill, 2011; Mulligan, 2017).

Designing and enabling opportunities for personally relevant transformative learning experiences is a major challenge for both our formal and informal educational systems (Hill, 2001; 2011; 2012; Hill et al., 2004; Sattmann-Frese & Hill 2008; Lange, 2009; Leal Filhoaj et al., 2018; Visser, 2018; Thomashow, 2020). The foundational task of an effective educational system is to enable learners to understand the breadth and depth of our situation and the future possibilities, and to develop the competencies needed to individually and collectively design and implement responsible and effective actions and systems that can meet our needs equitably and sustainably.

Below I have highlighted **four areas for inclusion in transformative educational programmes**. I consider that, to be effective, all learning experiences need to:

- be unique to the individual (taking into account content, time, place, modes of delivery and contextual relevance);
- nurture relationships (with the teachers – as mentors and guides; with other learners – collaborators in learning; and with supporters – including family and community members);
- develop wisdom and ethics (being), as well as knowledge and skills (doing);
- be supportive of the sort of goals that I am advocating here.

Democratic schools and *Big Picture Education* schools[10] are among those that exemplify much of what I am advocating.

Four priority areas of understanding for inclusion in curricula for personal and cultural transformation

Social ecology

Throughout history there have been many individuals and small groups who have sought to engage in divergent, holistic thinking, to challenge the *status quo*, to envision better futures for all and to work to implement them. Invariably they have suffered the consequences of doing this, most commonly through marginalisation and lack of understanding and support. These have been the experiences of many working within a social ecological framework, and also of those critiquing dominant approaches in education, including those promoting transformative learning (Christie et al., 2015).[11]

The qualities that distinguish my *social ecological framework and worldview* from other well-meaning approaches to the understanding of systems and change include the following:

Inclusion of a consideration of the **personal** (including our psychosocial histories and conditions) in all change initiatives, as well as social (including historical, political, economic, business and technological factors – without privileging any of these) and environmental and spiritual factors.

An acknowledgement of the vast **unknown** (and the miniscule known, the focus of most education, with its current naïve over-emphasis on STEM: Science, Technology, Engineering and Mathematics). Education needs to devote much more attention to enabling students acquire the competences for engaging with the unknown, through experience and reflection, and to the development of wisdom. This must avoid treating the unknown as if it is known, as is the practice within most religions.[12]

The need to develop an awareness of the factors influencing the **design** and management of all institutional structures and processes, and all thinking and action, taking into account critical issues relating to gender, power, and difference, and the need to develop competencies in wellbeing-enabling system design, collaboration (across difference) and respectful and caring communication.[13]

Because such a social ecology framework is supportive of developing a broad and deep understanding of the complex relationships between human behaviour and social and environmental processes, and the need for more inclusive, long-term improvement programmes, I consider that it is ideal for supporting the design and implementation of the curriculum and teaching methods that I am arguing for here. However, because of social ecology's inability to fit neatly into the narrow disciplines and silos that characterise most current educational institutions, most social ecology units and initiatives have been subject to ongoing erosion of their cultural capital, and to marginalisation and even elimination. Yet, paradoxically, what social ecological thinking has to offer is needed now more than ever to enable the necessary transition to an ecologically sustainable, meaning-rich, wellbeing-enabling society (Bookchin, 1962; 1971; 2015; Wright et al., 2011; Stokols, 2018).

Equitable wellbeing

I regard equitable wellbeing as the bottom line, not economics and the economy. Money and the economy must always be in the service of life, and supportive of the factors that enable life and wellbeing. Although money and economic processes can be useful and convenient aids in enabling us to act on our higher values, money and growth must never be regarded as a higher value (as they tend now to be). Globalisation has enlarged the magnitude of this challenge, such that most people now feel impotent in the face of the meaningful changes that are needed. Fortunately, recognition of this problem[14] has recently generated some bold fresh economic thinking.[15]

For me, the most profound insight into economics came from reading the "parable" by Nicholas Georgescu-Roegen (1971, p. 300), who noted that whereas the rate of germination of a grain of rice has changed little over time (in response to bio-ecological limits), rates of manufacturing in factories have increased by orders of magnitude. He argued that the economic systems that dominate today are essentially "factory economic systems" (mechanistic), whereas for agriculture (and other nature-based systems) we need "farm economic systems" (systems that are tuned to life rather than to machines). It seems to me that because people (who are living beings, like rice) work in factories, we actually need only "farm economics" for everything, and that it is "factory economics" that is helping to prevent us from relating responsibly to our support environment, and to one another.

Sustainability

Ensuring ecological sustainability for humans requires us to responsibly design and manage our being and doing in ways that take into account the same four interrelated factors that determine the sustainability of other species: our numbers, distribution, activities and the carrying capacity of the support environment. Our unsustainability increases with population density, distribution away from essential resources, high consumption and waste production, and (as it does for all other life) with declining carrying capacity of support environments.

All of these areas require urgent attention in terms of policy development and work practice, including those needed to support cultural and personal transformation.

Thus, we can achieve sustainability if we keep population density at replacement levels, live a conserver lifestyle (not engaged in "compensatory" consumption and growth), distribute ourselves, where possible, close to required resources and where wastes can be returned to natural systems or be totally recycled, and recognise that all activities must be subject to bio-ecological limits that respect the requirements of all life (avoiding novel chemical and other life-undermining products and practices). In particular, we need to realise that the most important activities in well-functioning ecosystems are concerned with *system maintenance*, with *sustainable production* being a by-product of well-maintained systems. It would be wise to design and manage all of our technologies and work activities with this in mind. For humans, this requires the building up and maintenance of personal (including spiritual), social and natural capital (our soil, fresh water and marine ecosystems). Our over-focus on production (and not system maintenance) is what is largely responsible for the increasing loss of capital and system breakdown in all of these areas. I have recently reflected on this more extensively in relation to our food systems, and discussed the design and psycho-social understandings and actions needed to enable meaningful change (Hill, 2014).

Leadership

My experience in the workplace is that most people labelled as leaders today are increasingly functioning primarily as managers. They do not realise that all effective management must function in the service of responsible and visionary leadership, which is values-based, and emphasises creativity, system design, enabling team collaboration, monitoring processes and outcomes, and responding appropriately to such feedback. At a foundational level, wise leadership is primarily concerned with *integration* (rather than separation and fragmentation), enabling contextual *balance* (rather than just control and efficiency), and attention to local to global, and short- to long-term, diverse expressions of *feedback* (rather than focusing only on naïve measures such as productivity, profit and power). This is as important, and as much a problem area, in our educational systems as in other areas. The lack of strong, enlightened leadership is why very little within current education programmes is concerned with the content of this chapter (and this book), including especially the development of wisdom (Ardelt, 2018; Bruya & Ardelt, 2018), and working towards enabling ecological sustainability, equitable wellbeing, meaning, inclusion and collaboration.[16]

Enabling meaningful change: including "deep" personal learning, development and action, and radical organisational and institutional transformation to enable this

When working with change, it is important to be clear about the essential nature of humans. All humans are social organisms[17] with immense abilities (and innate passions) to learn, create, relate, reflect and share. As we grow up, experiences (at home, in school and in society) that fail to adequately acknowledge and nurture these qualities, and that obstruct their expression, require us to adapt to the repetition of these experiences. This results, for most of us within industrialised societies prior to age 7 to 11, in the development of a range of *adapted selves* (in addition to our *core self*) as automatic survival strategies. Whereas one's core self is largely present-focused (contextual), spontaneous, loving, relational, empowered, aware and connected to one's forming systems of ethics and values, one's adapted selves are defensive and tend to be more connected to the past and future, exhibiting patterned behaviours (which range from withdrawal to over-expression), creating distorted constructions about ourselves, others and the world, being somewhat fearful and self-focused, to some extent disempowered, less aware, and more disconnected from one's ethics and values. Although these adaptive selves were originally valuable in enabling us to survive *oppressions*, their subconscious automatic persistent expression over the longer term means that they become increasingly maladaptive and harmful,[18] and significant barriers to transformative learning (and the needed personal-to-global transformation and responsible action).

The foundational key to being effective in enabling meaningful learning, development and change is to ensure that the only communication taking place is between the core self of teachers and facilitators and the core self of students and all others. All communication with adapted selves reinforces their existence and is a barrier to transformational learning and meaningful change and action. This partly explains the effectiveness of *democratic education* (Hecht, 2012) and *action learning* (Aubusson et al., 2009), as both require more equitable, contextually relevant engagement, and equitable collaboration, within learning environments than the more dominant didactic, hierarchical approaches.

Lying to change the world

Probably the most ground-breaking discovery I have made in my life is what I am about to share, and yet I am aware that it is likely to be the most difficult for others to embrace because of the psychological processes that I have discussed above. It is, however, the best example I have of doing what I am advocating here.

If I ask students what they would like to learn in my classes, what projects and research they would like to carry out, and what they would like to achieve in their lives, their responses, without being aware of it, will all have been subconsciously censored by their multiple *adaptive/maladaptive selves*. They say things that they sense will be acceptable within the present context, not what their *core selves* would really like to say, do and achieve.

By accident, I found that if I asked them to ***boldly lie*** about what they had already learned in my classes, what projects (and research) they had already successfully carried out, and what major things they had done in their lives that had enabled significant improvements – in their personal lives, families, groups, communities, at work, politically, environmentally, etc. – I experienced their uncensored *core selves* responding, and in powerful and deeply personally relevant ways. I recognised this as a paradoxical transformative process.

I then asked them to recall anything that they had actually done that had any relationship with their lies. All had, not surprisingly, already done some such things. Then I asked them what might be a next step from what they had done towards their lie(s), what they might need to enable them to take this step, how they might get the needed resources and supports, what might get in the way, and how they might get around those barriers (i.e., applying Lewin's (1935) Force Field Analysis).

Finally, I asked them to make a 100% commitment to a doable action (however small), and to identify who might be an ally in taking the action and what they would ask from them. I also asked them how they might *celebrate their successes* so that others might learn from them, thereby *making their small, meaningful initiatives contagious*.

Although there are dozens of other things that I could share from my experiences of enabling meaningful transformative learning and change,[19] my best suggestion is that you dare to do what I have described above, and also dare to enable all involved to celebrate the outcomes, so that the understandings and processes involved can be widely shared.

This is not only relevant for transforming our educational systems, but also our political, business, health, social services and all other systems. Go well on your transforming journeys!

Notes

1 Since I started writing in the 1960s I have imagined that what I am writing may be both my first and last piece of writing: first in terms of saying new things, and last in terms of not holding anything back. I have also been fortunate throughout my life for having been supported in my work by courageous mentors and colleagues, to whom I will always be grateful.

2 Nafeez Ahmed, 23 Nov 2019. The Collapse of Civilization May Have Already Begun. Vice (https://www.vice.com/en_us/article/8xwygg/the-collapse-of-civilization-may-have-already-begun); a helpful review of the diverse responses to the "collapse" ideas.

3 A 2013 survey of Australians found, for example, that "9 in 10 regard themselves as without influence over the federal level of government" (Evans et al., 2013).

4 I have critiqued this coping behaviour as monitoring our extinction research, which is commonly prioritised and funded by those with inequitable personal and institutional power to postpone and avoid responsible action, and so be able to retain their power. It is a significant "sink" for research and social change funding that could be better used to address causes of problems and enable meaningful change (Hill, 2014).

5 A wise medical friend put it most clearly when he commented that "we don't suffer from headaches because of a deficiency of Aspirin in the blood."

6 Four recent texts that I think have made significant contributions to rethinking and redesigning our cultures are Hamilton (2017), Dale (2018), Washington and Twomey (2018) and McKibben (2019).

7 Other useful texts include Flannery (2016) and Orr (2016).

8 A class-handout I prepared to support this learning may be obtained by writing to me (s.hill@westernsydney.edu.au).

9 I define social ecology as: "the study and practice of personal, social and ecological sustainability and 'progressive' change based on the critical application and integration of ecological, humanistic, relational, community and 'spiritual' (the unknown) values to enable sustained wellbeing of all" (slightly updated from Hill 2011). This reflects key elements of the Social Ecology approach developed since the 1980s at what is now Western Sydney University. Our project- and presentation-based programmes (no exams) aimed to develop student's communication and professional skills, together with three interrelated meta-skills: 1. Understanding and working with complexity, power, gender and otherness/difference, 2. Exploring diverse ways of knowing and 3. Creative visioning and designing.

10 http://www.educationrevolution.org/store/findaschool/democraticschools/; https://www.bigpicture.org.au/. However, because we learn in school from those individuals who are able to inspire and enable meaningful and relevant learning, I have always advised my university students to choose teachers rather than subjects. The same logic applies to schools and the teachers who work in them. So, the name of a school is no guarantee of the quality of all of its teachers.

11 Transformative learning is learning that enables irreversible, profound, emancipatory change for the better – in our values, world views, beliefs, perspectives, understandings, and frameworks (meaning schemes) for imagining, thinking, designing, planning and acting, as well as in our day-to-day living and relating (to self, others, and the built and natural world). It is the "highest" level of learning: above "refining or elaborating our meaning schemes, learning new meaning schemes, [and] transforming meaning schemes" (Mezirow, 1994, p. 224).

12 This paradoxically is a barrier to the deep, expansive experiences needed to relate humbly, meaningfully and responsibly to one another, other species, and the planet.
13 I have listed key ecological design literature in Hill (2014, p. 414). Other key recent references are Van der Ryn (2013) and Todd (2019).
14 I was originally alerted to these issues through the inspirational writings of Ernst Schumacher (1973), Tom Bender (1975), Hazel Henderson (1978), and Herman Daly (1980).
15 Recent bold economic analyses and initiatives have been provided by Milani (2000); Eisler (2007); Smith & Max-Neef (2011); Lietaer & Dunne (2013); Mitchell et al. (2016); Bregman (2017); Monbiot (2017); Hinton & Maclurcan (2018); Raworth (2018); and Trainer (2019).
16 Relatively few texts on leadership are compatible with the "deep and broad" approach that I am advocating here. These eight texts include some valuable insights: Spears & Lawrence (2001); Wren (2007); Kay & Venner (2010); Marshall et al. (2011); George (2015); Brown (2018); Hawkins (2018); and Storm & Hutchins (2019).
17 So, we don't need to be "socialised"; but we do need to be enabled to fully express and develop our social potential, which has profound implications for the structure and functioning of our parenting and educational systems. This was most clearly illustrated in the Peckham Experiment (www.thephf.org) (Stallibrass, 1989).
18 I have listed what I regard as the key psychological literature, relevant to this discussion, which reflects my particular understanding of human psychology, in a previous publication (Hill, 2003, p. 189); see also Somerville (2018).
19 Other suggestions for enabling meaningful change can be found in my many PowerPoint presentations (www.stuartbhill.com) and in my publications. Some of these presentations and publications, and the handouts for my workshops on "Lying to change the world," may be downloaded from: https://www.researchgate.net/profile/Stuart_Hill6.

References

Ardelt, M. (2018). Can wisdom and psychosocial growth be learned in university courses? *Journal of Moral Education*, 40(1), 30–45.

Aubusson, P., Ewing, R., & Hoban, G. (2009). *Action learning in schools: Reframing teachers' professional learning and development*. Routledge.

Bendell, J. (2018). *Deep adaptation: A map for navigating climate tragedy: Institute for Leadership and Sustainability Occasional Paper 2*. http://lifeworth.com/deepadaptation.pdf

Bender, T. (1975). *Sharing smaller pies*. Rain.

Bookchin, M. (1962). *Our synthetic environment*. Albert A. Knopf.

Bookchin, M. (1971). *Post-scarcity anarchism*, Ramparts.

Bookchin, M. (2015). *The next revolution: Popular assemblies and the promise of direct democracy*. Verso.

Bregman, R. (2017). *Utopia for realists: And how we can get there*. Bloomsbury.

Brown, B. (2018). *Dare to lead*. Random House.

Bruya, B., & Ardelt. M. (2018). Fostering wisdom in the classroom, part 1: A general theory of wisdom pedagogy. *Teaching Philosophy*, 41(3): 239–253.

Christie, M., Carey, M., Robertson, A., & Grainger, P. (2015). Putting transformative learning theory into practice. *Australian Journal of Adult Learning*, 55(1), 10–30.

Dale, A. (2018). *Edging forward: Achieving sustainable community development*. Fernweh.

Daly, H. E. (1980). *Economics, ecology, ethics: Essays toward a steady-state economy*. W. H. Freeman.

Eisler, R. (2007). *The real wealth of nations: Creating a caring economics*. Berrett-Koehler.

Evans, M., Stoker, G., & Nasir, J. (2013). *How do Australians imagine their democracy? Australian survey of political engagement findings 2013*. ANZSOG Institute for Governance at the University of Canberra.
Flannery, T. (2016). *Atmosphere of hope: Searching for solutions to the climate crisis*. Text Publishing.
Fletcher, R. (2016). Neoliberalism and environmental education. In M. A. Peters (Ed.), *Encyclopedia of educational philosophy and theory* (pp. 1–5). Springer Science and Business Media.
George, B. (2015). *Discover your true north* (2nd ed.). John Wiley & Sons.
Georgescu-Roegen, N. (1971). *The entropy law and the economic process*. Harvard University Press.
Greenleaf, R. K. (2002). *Servant leadership: A journey into the nature of legitimate power and greatness*. Paulist Press. (Original work published 1977).
Greenwood, D. (2010). A critical analysis of sustainability education in schooling's bureaucracy: Barriers and small openings in teacher education. *Teacher Education Quarterly*, 37(4), 139–154.
Hamilton, C. (2017). *Defiant Earth: The fate of humans in the Anthropocene*. Allen & Unwin.
Hawkins, P. (Ed.). (2018). *Leadership team coaching in practice: Case studies on developing high-performing teams*(2nd ed.). Kogan Page.
Hecht, Y. (2012). *Democratic education: A beginning of a story*. Innovation Culture.
Henderson, H. (1978). *Creating alternative futures: The end of economics*. Berkley Pub. Corp.
Henderson, J., Hursh, D., & Greenwood, D. (Eds.). (2018). *Neoliberalism and environmental education*. Routledge.
Hill, S. B. (2001). Transformative outdoor education for healthy communities within sustainable environments. In *12th National Outdoor Education Conference: Education Outdoors – Our Sense of Place – Conference Proceedings* (pp. 7–19). Victorian Outdoor Education Association.
Hill, S. B. (2003). Autonomy, mutualistic relationships, sense of place and conscious caring: A hopeful view of the present and future. In J. I. Cameron (Ed.), *Changing places: Re-imagining Australia* (pp. 180–196). Longueville.
Hill, S. B. (2011). Social ecology: An Australian perspective. In D. Wright, C. Camden-Pratt, & S. B. Hill (Eds.), *Social ecology: Applying ecological understanding to our lives and our planet* (pp. 17–30). Hawthorn.
Hill, S. B. (2012). Education to change the world: learning for realising one's personal, social & ecological potential. *21st Century Learning*. https://www.researchgate.net/publication/302906192_Education_to_Change_the_World_Learning_for_Realising_One's_Personal_Social_Ecological_Potential
Hill, S. B. (2014). Considerations for enabling the ecological redesign of organic and conventional agriculture: A social ecology and psychological perspective. In S. Bellon & S. Penvern (Eds.), *Organic farming: Prototype for sustainable agricultures* (pp. 401–422). Springer.
Hill, S. B., Wilson, S., & Watson, K. (2004). Learning ecology: A new approach to learning and transforming ecological consciousness: experiences from social ecology in Australia. In E. V. O'Sullivan & M. Taylor (Eds.), *Learning toward an ecological consciousness: Selected transformative practices* (pp. 47–64). Palgrave Macmillan.
Hinton, J., & Maclurcan, D. (2018). *How on Earth: Flourishing in a not-for-profit world by 2050*. Post Growth Publishing.
Hunter, D., Bailey, A., & Taylor, B. (1997). *Co-operacy: A new way of being at work*, Tandem.
Kay, S., & Venner, K. (Eds.). (2011). *A cultural leadership reader*. Creative Choices.
Kingsolver, B. (2012). *Flight behavior: A novel*. Harper.
Klein, N. (2014). *This changes everything: Capitalism vs. the climate*. Simon & Schuster.

Laing, R. D. (1971). *The politics of the family*. Penguin.

Lange, E. A. (2009). Fostering a learning sanctuary for transformation in sustainability education. In J. Mezirow, E. W. Taylor (Eds.), *Transformative learning in practice: Insights from community, workplace, and higher education* (pp. 193–204). Jossey-Bass.

Leal Filhoaj, W., Raathb, S., Lazzarinic, B., Vargasd, V. R., de Souzab, L., Anholone, R., Quelhasf, O. L. G., Haddadg, R., Klavinsh M., & Orlovici, V. L. (2018). The role of transformation in learning and education for sustainability. *Journal of Cleaner Production* 199, 286–295.

Lietaer, B., & Dunne, J. (2013). *Rethinking money: How new currencies turn scarcity into prosperity*. Berrett-Koehler Publ.

Lewin, K. (1935). *A dynamic theory of personality: Selected papers by Kurt Lewin*. McGraw Hill.

McKibben, B. (2019). *Falter: Has the human game begun to play itself out?*Henry Holt.

Marshall, J., Coleman, G., & Reason, P. (Eds.). (2011). *Leadership for sustainability: An action research approach*. Routledge.

Mezirow J. (1994). Understanding transformation theory. *Adult Education Quarterly*, 44(4): 222–232.

Milani, B. (2000). *Designing the green economy: The postindustrial alternative to corporate globalization*. Rowman & Littlefield.

Mitchell, W. F., Wray, L. R., & Watts, M. J. (2016). *Modern monetary theory and practice: An introductory text*. CreateSpace Independent Publishing Platform.

Monbiot, G. (2017). *Out of the wreckage: A new politics for an age of crisis*. Verso Books.

Mulligan, M. (2017). *An introduction to sustainability: Environmental, social and personal perspectives* (2nd ed.). Routledge.

Orr, D. (2016). *Dangerous years: Climate change, the long emergency, and the way forward*. Yale University Press.

Quinn, D. (1992). *Ishmael*. Bantam.

Raworth, K. (2018). *Doughnut economics: Seven ways to think like a 21st-century economist*, Penguin.

Read, R., & Alexander, S. (2019). *This civilisation is finished: Conversations on the end of empire – and what lies beyond*. Simplicity Institute Publishing.

Sattmann-Frese, W., & Hill, S. B. (2008). *Learning for sustainability: Psychology of ecological transformation*. Lulu.

Schumacher, E. F. (1973). *Small is beautiful: A study of economics as if people mattered*. Sphere Books.

Smith, P. B., & Max-Neef, M. (2011). *Economics unmasked: From power and greed to compassion and the common good*. UIT Cambridge.

Somerville, W. (2018). *Shallow thought deep mind: What you need to succeed, thrive and make a better world*. Dr Wayne Somerville.

Spears, L. C., & Lawrence, M. (Eds.). (2001). *Focus on leadership: Servant-leadership for the twenty-first century*. John Wiley & Sons.

Stallibrass, A. (1989). *Being me and also us: Lessons from the Peckham experiment*. Scottish Academic.

Stokols, D. (2018). *Social ecology in the digital age: Solving complex problems in a globalized world*. Academic Press.

Storm, L., & Hutchins, G. (2019). *Regenerative leadership: The DNA of life-affirming 21st century organizations*. Wordzworth Publ.

Thomashow, M. (2020). *To know the world: Why environmental learning matters*. MIT Press.

Todd, J. (2019). *Healing Earth: An ecologist's journey of innovation and environmental stewardship*. North Atlantic Books.

Trainer, T. (2019). The "Nobel Prize" for Economics 2019… illustrates the nature and inadequacy of conventional economics. *Real-World Economics Review*, 90, 41–46.

Van der Ryn, S. (2013). *Design for an empathic world: Reconnecting people, nature, and self.* Washington DC: Island Press.

Visser, J. (2018). *The Anthropocene: A different learning landscape for a different world.* [Lecture delivered at the Universitas Nusa Cendana]. Learning Development Institute. www.lea rndev.org/dl/Kupang2018.pdf

Washington H., & Twomey, P. (Eds.). (2018). *A future beyond growth: Towards a steady state economy*. Routledge.

Wren, J. T. (2007). *Inventing leadership: The challenge of democracy*. Edward Elgar.

Wright, D., Camden-Pratt, C., & Hill, S. B. (Eds.). (2011). *Social ecology: Applying ecological understanding to our lives and our planet.* Hawthorn.

4

WHAT WAS EDUCATION FOR?

Learning in the shadow of climate change

Isak Stoddard

> "Is education possibly a process of trading awareness for things of lesser worth?"
> – *Aldo Leopold* [1]

The latest science on climate change, and the increasingly rapid obliteration of the more-than-human world,[2] calls for an uncomfortable and fundamental shift in our thinking and practice. What we have tried so far, at universities and in society at large, has utterly failed to curb emissions and to stop the continued fragmentation of ecological communities across the globe. Furthermore, many of us privileged enough to find ourselves at a university are deeply implicated – both indirectly and more directly[3] – in the destruction of ecological life-support systems for humans and other life on Earth. For many of us, breaking our routines and patterns of behaviour is what is needed: a process of unlearning, where we lose our foothold and become more fully aware of our predicament. Perhaps then, unexpected insights on how to respond and where to go next may emerge.

Higher education and the reproduction of climate change

Most modern educational systems, including primary and secondary school, but perhaps most prominently universities and higher education, are deeply implicated in creating and maintaining the structures, mindsets and lifestyles that are driving and exacerbating climate change and other associated challenges to ecological sustainability. As David Orr (1991) points out in his seminal essay "What is Education For?" it is not the work of ignorant people that is destroying the world:

> The truth is that many things on which your future health and prosperity depend are in dire jeopardy: climate stability, the resilience and productivity of natural systems, the beauty of the natural world, and biological diversity. It is

> worth noting that this is not the work of ignorant people. It is, rather, largely the result of work by people with BAs, BSs, LLBs, MBAs, and PhDs.
>
> *(p. 1)*

The so-called environmental crisis can also be seen as a fundamentally human crisis. "We are the environmental crisis," as philosopher Neil Evernden (1993) has said. Following this line of thought, it is not such a giant leap to ask if perhaps our institutions of higher education are also in a form of crisis, an organisational as well as existential crisis that mirrors and reproduces the ecological crisis that is unfolding around us this very minute. How might we as students, educators, academics and concerned inhabitants of Earth respond to the perilous situation that we find ourselves in? What forms of learning may guide us best in these stormy and uncertain times in which we live? In what ways could we start to think differently about education, within as well as outside of formal educational institutions? For us to respond to these questions, I believe it may be wise to pause for a moment, and to first face how dire and urgent the situation actually is.

The long shadow of climate change

2016 was the warmest year in recorded history, 2019 the second warmest year (WMO, 2020). Runners-up are 2015, 2017 and 2018 (WMO, 2019). Six of the warmest years ever have taken place since 2010 (NOAA, 2018). For many humans and other life forms on this planet, the ecological consequences of climate change are already making life difficult and increasingly a threat to their existence.[4] Even in a seemingly "insulated" country such as Sweden, the summer of 2018 was something beyond normal, with raging forest fires and wells running dry across the country. Events such as these are likely to get much worse as global warming really kicks in. We are living in the long shadow of climate change.

The Paris Agreement on climate change was a triumph in that it established a universal commitment amongst world leaders to take action to "hold the increase in global average temperature to well below 2°C above pre-industrial levels and pursue efforts to limit the temperature increase to 1.5°C." But, as environmental writer George Monbiot (2015) wrote the day after the Paris Agreement was negotiated: "By comparison to what it could have been, it's a miracle. By comparison to what it should have been, it's a disaster."

Despite decades of optimistic rhetoric, global emissions have risen over 60% since the Intergovernmental Panel on Climate Change (IPCC) published its first report in 1990. Since the Paris Agreement was adopted in 2015, global emissions have continued to rise. The latest IPCC special report on global warming (IPCC, 2018) calls for immediate and transformative action for us to have a chance of avoiding catastrophic climate change, but unfortunately then fails to address the profound implications of reducing emissions in line with the temperature commitments enshrined in the Paris Agreement (Anderson, 2018a). Scientists and modellers grappling with describing viable ways forward are now conjuring up (and almost without exception relying on) spurious so-called "negative emission technologies" to make the numbers work (Anderson & Peters, 2016).[5] They seldom acknowledge that climate change now means whole systems change, either of the involuntary kind driven by runaway climate change or by an immediate and direct challenge to the dominant socio-economic paradigm, its extractive industries and the mantra of green growth.[6] Worse still, instead of focusing on decreasing emissions today, some seem to be using the climate crisis to suggest that the intentional manipulation of the Earth systems at a massive scale, known as geoengineering, may become necessary in the near future.[7]

The reality of climate change is already upon us and about to get much worse. There are, however, great risks involved with conceptualising climate change as a "problem to be fixed." Climate change is a symptom of an utterly unsustainable and destructive form of human civilisation, and to a very large extent embedded in the structures and practices of industrial societies, including the momentum and vested interests of industry and consumers still dependent on the extraction of fossil fuels. That said, we cannot ignore the fact that the physical drivers of anthropogenic climate change are fairly simple and straightforward. The amount of emissions released, primarily through the combustion of fossil fuels, is what is heating the planet. This is captured succinctly in the scientific framing of carbon budgets, which gives us the total amount of cumulative emissions that can be released to stay below a certain temperature threshold. If we are to deliver on the temperature commitments we have made in the Paris Agreement, we only have around 15 years of current global emissions left before we have spent our remaining carbon budget.[8] Industrialised, wealthy nations would need to decrease their carbon emissions by 10% per year as an absolute minimum, starting immediately (Anderson et al., 2020). In this sense, addressing climate change is more straightforward than responding to other environmental and sustainability issues such as biodiversity loss, species extinction or rapidly increasing socio-economic inequalities.[9] To avoid the most catastrophic climate scenarios we simply need to decrease global emissions very rapidly.

But who is "we"? The sources of global emissions are highly skewed. Those most likely to suffer the worst consequences of climate change (at least initially) are those least responsible for driving both historic and contemporary emissions skyward. Almost fifty per cent of global emissions come from the activities of just 10% of the global population (Chancel & Piketty, 2015). Also, those of us with the greatest political and economic power in our societies (and high-carbon lifestyles) tend to be from a generation that won't be around when the worst feedbacks and dynamics of climate change may start.[10]

If the top 10% of global emitters would decrease their carbon footprint to the level of the average European, we would see a 1/3 cut in global emissions (Anderson, 2018b). Those of us in this privileged top 10% might see what needs to be done, but do we have the courage and wherewithal to really push for a truly progressive agenda for addressing climate change, knowing that it would impose limits (both real and perceived) on the way we currently live our lives?

There is also an interspecies equity dimension to climate change that often is overlooked. Humans are not only directly responsible for the ongoing mass extinction of species, but also the main culprit in the unfolding drama of climate change, with impacts that will first and foremost affect those least responsible for causing them, in this case, other-than-human life on Earth.

Climate change is fundamentally about equity: a question of intergenerational, intragenerational and interspecies justice. The philosopher Stephen Gardiner has called it a perfect moral storm, where the global and intergenerational nature of climate change is further complicated by our ineptitude in thinking about and acting upon problems of the long-term future (Gardiner, 2011). Could we still envisage forms of higher education that would embody a meaningful and wise response to the predicament we find ourselves in?

Higher education for liveable climate futures

What might be meaningful and appropriate responses to the dire state of both our climate and our institutions of higher education? What are we called to do?

I don't know yet. But I have been pondering and working with these questions ever since I first came into contact with the student-initiated and transdisciplinary Centre for Environment and Development Studies (CEMUS) at Uppsala University, and the Swedish University of Agricultural Sciences, back in 2005. Over the past 15 years I have had the privilege of studying and working within the collaborative environment at CEMUS in a number of different capacities. Over these years we have tried, and engaged with others who have tried,[11] to bring about a transformation of higher education and universities in response to the most pressing social and ecological issues of our times. At times, it has been immensely rewarding and inspiring, and at other times equally frustrating, even bordering on despair.

During an especially frustrating period of work at the university a few years ago I came across the paper *Mapping Interpretations of Decolonization in Higher Education* (Andreotti et al., 2015) by the Canadian-Guarani scholar Vanessa Andreotti and her colleagues. As a pedagogic exercise, they introduce a social cartography of various responses to engage with critiques of modernity within the context of higher education. The set of responses include pretending that there is no problem ("everything is awesome"), "soft reform" responses, "radical reform" responses, and onto "beyond reform."[12] While reading the paper, I found myself especially drawn to the two latter categories, recognising approaches that I also have explored when working to enable transformation within the university. Interestingly, "beyond reform" did not necessarily imply giving up – to pack up and walk out. *Developing alternatives, hacking the system* and *hospicing* were identified as possible alternatives to quitting the game and walking away from an institution that you have come to see as "beyond reform."

What now follows are a few remarks stemming from my own experiences working for change within Uppsala's two universities, and an underlying commitment to working towards a more sustainable and just future. The remarks are perhaps best understood as the beginning of a conversation rather than an action plan or recommendations, and in the social cartography introduced above, they weave in and out of the soft and radical reform responses, with a few visits into the beyond reform space. My hope is that they can be of benefit, and might act as a source of inspiration (or solace), for students, educators and anyone else concerned with climate change, the ecological crisis and the future of higher education.

Ostrich or a phoenix? On the transformation of universities

Will universities act as an ostrich, sticking its inquisitive head deeper into the sand as social and ecological crises are mounting? Or will they act as the mythical phoenix, reborn and finding new and meaningful forms in the flickering shadows of climate change?[13] Which forces and interests should shape their future direction?

One of the central ideas that grew out of the creation of CEMUS, back in 1992, was the idea of intergenerational justice. If the social and ecological consequences of our actions today will have to be dealt with by the younger generation, then they should have the opportunity to shape the conditions, content and form of their education, so they may have better chances of being able to deal with this uncertain and challenging future. Universities, in their present state, are almost never designed for those ends (Stoddard et al., 2012). The simple yet revolutionary idea at CEMUS is to have students themselves lead this inquiry, with the acquired support of the intellectual community, organisational resources and societal platforms that a university has to offer. It is a space for students to propose, develop and deliver course content and forms of education that aim to bridge and constructively challenge the organisational, cultural and conceptual divisions at a university – between disciplines, between teacher and student and between university and society (Hald, 2011).[14]

Although CEMUS was initiated – and continues to be shaped by – students, the contributions and support from senior academics (researchers and faculty) have always been a crucial part of the CEMUS educational model. Conversely, CEMUS' learning environment has also been a crucial part of many researchers' and faculty members' academic life. A former student and course coordinator at CEMUS wrote a thesis about transformations at universities in which he described CEMUS as a "shadow space for social learning, a community of practice, semi-detached from institutional context, in which innovative capacities for meeting the challenges of implementing sustainable development at universities has been built up and nurtured over time" (Anderson, 2017, p. iii). Unique capacities of CEMUS identified by Anderson (2017) include new forms of student–faculty partnerships, working across and between disciplines as a matter of practice and, perhaps most importantly, re-purposing and redesigning university education to create a knowledge base for learning and action on urgent social and ecological challenges. I am becoming increasingly convinced that at the heart of the idea of CEMUS also lies a new model of collaborative knowledge creation and community building that elucidates important research questions and practices for understanding and empathetically responding to the mess the world is in.

CEMUS is an example of how a group that will be strongly affected by the consequences of climate change (the younger generation) is able to propose and deliver course content and pedagogy of university courses and programmes. Perhaps one could take this idea and practice a bit further, allowing voices and perspectives of some of the other strongly affected groups to shape the curriculum.[15] There is a saying in Brazil: "In a situation of a flood, it's only when the water reaches your bum that you can actually swim."[16] What forms could higher education take if people with "water up to their bums" in our societies (be it from climate change or other social, economic or ecological pressures) would provide insights that directed how and what we do at universities in response to climate change?[17]

But what about species other than humans? If they had a say, what would a bee or a bear want us to study and do at university? What would the curriculum look like? What sort of learning would take place if students of economics, engineering, law and political science would be guided by the interests of the bees?[18] To some, this may seem like an odd or unreasonable question, but considering our utter dependence on pollinators[19] such as bees to provide humanity with food, it may not be such an irrelevant question after all. We need to find ways to introduce more eco-centric approaches into anthropocentrically organised societies and knowledge structures. Or perhaps wiser yet, find ways to reclaim anthropocentrism by discarding the false dichotomy between nature and culture, and positioning humanity once again at the heart of an animate and living world.

Finally, what would the particular places and bioregions surrounding our institutions of higher education want us to study? How could we find out? What stories and opportunities for new forms of learning exist just outside the classrooms and lecture halls? How could we begin exploring the idea of a place-driven education and curriculum development by unearthing hidden or forgotten histories of the land, studying social and ecological relationships through face-to-place storytelling and walking? Pedagogic experiments could revive the ancient Aristotelian tradition of peripatetic walks, or draw inspiration from contemporaries such as the philosopher and educator Jan Masschelein, where walking along lines in the city becomes the starting point for a learning journey (Masschelein, 2010). In developing such ecologically and locally informed curriculums, approaches from social ecology (e.g. Wright & Camden-Pratt et al., 2011) have much to offer. Not least the concept of *learning ecology*, where ecological thinking and frameworks are introduced to better understand the process of learning while also integrating ecological understandings into the content of learning activities (Hill et al., 2004). What interesting and meaningful educational and organisational outcomes could arise when juxtaposing such a local and ecological context with the global perspectives required to address issues such as that of climate change and the "universal" ambitions of the university?

Over the past 28 years, over 15000 students, hundreds of teachers, researchers and others have contributed to the learning environment at CEMUS. Certainly, this has had a lasting impact on society, with many empowered and concerned students now engaged in work that builds on the insights and learning of having studied or worked at CEMUS. Nonetheless, no more than around 2–3% of the student population at Uppsala University end up taking a course at CEMUS. Although CEMUS has voiced a call for renewal at the university, financial and organisational limitations within a siloed university – including forces of institutionalisation and non-diversified funding streams –

have challenged CEMUS throughout its existence. In a blog piece I wrote for the 25th anniversary of CEMUS I compared the evolution of CEMUS and the university with that of a tree:

> A question that remains is if [CEMUS] is a smaller tree, shaded by the older trees above, with our students and roots holding CEMUS to the ground, and with branches and leaves reaching for the little sunlight that is let through the canopy. Or is CEMUS rather that unruly quality of life in trees, which might make branches turn in new and unexpected directions, refusing to conform to dominant trajectories set forth by external or internal circumstances?
>
> *(Stoddard, 2017a)*

Reflecting back, I would say we are both. But what distresses me, driven largely by the urgency of climate change, is that we really are in a hurry. What we are currently doing won't do, neither within our own institution and more broadly within universities. We hope, and have seen, that students and teachers in other parts of the world have been inspired and encouraged by our example, and that they continue to develop their local variations of CEMUS within their institutional and cultural contexts (Stoddard, 2017b). But there is also a need to reflect and further analyse the resistance we have met within our university and understand where it comes from. Perhaps this can give us insight into the resistance that can be found elsewhere in society when trying to bring about rapid and fundamental societal change. This brings me to a second set of reflections on the future of higher education for liveable climate futures.

The politics of learning and the transformation of society

Historically, students have played a key role in the creation of social movements that challenge the current political and economic order: from the student revolts beginning in Paris in 1968, to the Burmese protests of 1988, and the Arab spring of 2011.[20]

When it comes to climate change, the urgency of the situation requires immediate action, and a swift transformation of industrialised societies to fossil-free futures. Students and youth across the world are again at the forefront of raising important and difficult questions, demanding change and engaging in direct action to raise awareness around the incompatibility of the fossil fuel industry's operations with an appropriate response to the existential threat of climate change.[21]

These initiatives are surely all driving some form of change, and are playing an important role in bringing the issue of climate change into the public debate. However, in order to have a lasting effect on the organisation and content of higher education, these initiatives must be met with institutional responses from universities, with the active engagement of academics who have an understanding of the possible implications of climate change. So far, academics, including those researching climate change, have tended towards silent acquiescence, be it from a lack of insight and interest, or from fear of challenging the current economic and political paradigm (and research funding streams) – even when it comes at the cost of academic integrity.

There are small but promising indications that this might be beginning to change. More and more academics are beginning to organise to exert pressure on their peers and institutions to exhibit leadership on climate change – calling for an integration of climate change and other socio-ecological issues into teaching and learning across the university whilst showing that we believe our own research and rhetoric by rapidly reducing emissions from our own operations and activities, divesting from fossil fuels and acknowledging that we are in a climate emergency.

Although climate change is a matter of utmost urgency, there is a critical need to keep a broader and more long-term perspective in mind and to make sure that immediate, short-term actions taken do not dig humanity (and universities) deeper into the hole we need to climb out of. Universities have a responsibility and unique potential to convene different publics (Facer, 2020), to initiate conversations and healthy debates and enable new ideas on how to respond to climate change to emerge. Such initiatives would need to go well beyond the task of the university to merely communicate research results. It would require the university to dare to ask difficult questions about its implication in the reproduction of climate change, and to create platforms where relevant knowledge can be co-produced with citizens and civil society (Facer, 2018). In an increasingly digital age, and with climate change exerting an increasing amount of pressures on societies, physical meeting spaces for collaboration – as well as friction and disagreement – may become more and more important and needed.[22]

Unfortunately, there is a significant risk that the university as in institution will struggle to respond to the socio-ecological crisis in meaningful ways.[23] Rather than being called back to their civic purpose of being contributors to the common good, we may instead see universities continuing down the path of corporatisation and marketisation of higher education. A movement which started back in 2016, *Reclaim the university*, [24] is one of several recent attempts to bring attention to this dire development. Although initiatives such as these often are at the periphery, to me they indicate that there is indeed a deep frustration among many academics and students regarding the direction our institutions are taking, and that it befits us to at least try to do something about it.

And if our attempts fail (which they might very well do), or things get even worse (which is quite likely), then there is perhaps a point where we need to consider whether our efforts and capacities are better spent elsewhere, outside the walls of the university. Perhaps there may even be a critical moment where a significant number of academics and students concerned about climate change and other socio-ecological crisis no longer see much value or meaning in remaining within many of the formal institutions of higher education – that they are beyond reform. What new forms of higher education and communities of learning could then emerge?

Hope beyond hope

In many of the academic and public contexts I have found myself in lately, there has been a lot of talk about hope. Is hope a prerequisite for action? Or could it actually be what is keeping us from becoming more fully aware of our predicament and acting wisely? Perhaps hope at times works as an anaesthesia that numbs us from the ongoing ecological unravelling and heart-breaking loss of life around us. What happens if we let go of the blind hope that we can save our current unsustainable civilization? What would we gain from losing such hope? Maybe we would learn to acknowledge that loss is a part of life, to grieve for that which we care for but could not save, to fight for and protect that which we cannot bear to lose, and to celebrate the loss of that which we actually are better off without. So, is there a hope beyond hope?[25] I believe there is. As long as we acknowledge that "the end of the world as we know it is not the end of the world, full stop" (Kingsnorth & Hine, 2009).

This brings us again into the terrain of contemplating the space of "beyond reform." If a system, such as a modern, industrialised society, or a modern industrialised university, is falling apart, how can we assist in such a process in ways that create fertile ground for new sustainable and life-affirming beginnings?

These questions have been at the core of a strand of work I've been engaged with over the past ten years. It has found various forms, including a series of conferences around the theme of ClimateExistence, which brings together scholars, artists, students, activists and civil society to "unsettle some of the frames, structures and assumptions that shape our understanding of the world. Not necessarily to escape existing ones, but rather to explore the cracks where there is potential for alternative patterns to emerge." With these gatherings we have hoped to "collectively draw maps that inspire clarity, wisdom and imagination for humanity to be able to choose more humble and safe trajectories."[26] This line of inquiry has also informed many of the exploratory projects we have initiated with artists and artistic institutions across Sweden.[27]

This sort of work can often lead to contradictions, and a sense of being in a double or triple bind. What I have found so far is that an important step is to acknowledge that we are at a loss as to what to do, that what we have tried so far has not gotten us significantly closer to addressing the underlying issues that now are pushing us closer and closer towards the abyss of climate chaos and a full ecological unravelling. But perhaps, if enough of us admit that we are lost and begin to pay more attention to what is around us, we might yet find both bearable and humbling pathways for humanity in the years to come. In the words of an old Native American elder story, rendered into modern English by David Wagoner (1996):

If what a tree or a bush does is lost on you,
You are surely lost. Stand still.
The forest knows Where you are.
You must let it find you.

What was education for?

Looking back at our times, future generations might come to ask questions about the roles that educational institutions played when climate change went from being dire warnings in UN reports to a matter of life and death across the globe. One can only hope that they will be able to say, and truthfully so, that universities played a pivotal part in an unprecedented effort to enable a profound transformation of unsustainable societies, instead of continuing to justify them and their own role within them. One can only hope that they will be able to tell wonderful stories of students and academics around the world that had the courage to acknowledge the ways in which they were entangled in the mess the world is in, and that went on to find creative and humble ways of responding. One also hopes, echoing the spirit of Aldo Leopold, that they could describe how new forms of education brought awareness to many people about the things that truly matter in life, rather than becoming a fine-tuned social technology for trading awareness for things of lesser worth. And finally, one hopes that those with water up to their bums, those who have realised how to swim, will not be met with fear or contempt, but rather compassion and curiosity from people ready to listen and learn.

Notes

1 From A Sand County Almanac by Aldo Leopold (1949)
2 This term was first introduced by David Abram (1996) in his book *The Spell of the Sensuous: Perception and Language in a More-Than-Human World.*
3 Be it through fossil fuel emissions from transporting ourselves around the globe, from ecological consequences of our consumption of goods, from your investments, or from a silent acquiescence of an economic system and associated political agendas that are systematically undermining the wellbeing of ecologies and communities around the globe.
4 Exacerbating the already horrendous extinction rates we see today where global wildlife populations have fallen by 60% in just over four decades because of accelerating pollution, deforestation, climate change, and other anthropogenic factors (WWF, 2018).
5 For an illustrative example, see the recent paper "*Trajectories of the Earth System in the Anthropocene*" by Steffen, Rockström et al. (2018).
6 The scale of the decoupling of fossil fuel emissions from economic growth necessary to, for example, deliver on the Paris commitments, is now far too large to make this a viable strategy. See e.g. Hickel & Kallis (2019) and Parrique & Barth et al. (2019).
7 For an overview of current geoengineering technologies being researched, tested and developed, see http://www.geoengineeringmonitor.org/ and Wetter & Zundel (2018). Duncan McLaren's research also suggests that geoengineering modelling presumptions and practices may help deter mitigation and sustain elite interests in carbon-intensive economies (McLaren, 2018).
8 Based on global carbon budgets calculated in IPPCs special report *Global warming of 1.5° C* (IPCC, 2018) and emissions data from the Global Carbon Project www.globalcarbonproject.org.
9 However, ensuring that carbon sinks are not further eroded (e.g. through deforestation) and improved (e.g. through changing agricultural and other land-use practices) are also key to ensuring a stable climate and may also have many other ecological and societal benefits.
10 Our children will be around though, which begs the question: do we really care about our children?

11 Schumacher College (UK), The University of Bergen Collaboratory (Norway), the Social Ecology Group at Western Sydney University (Australia), the Green Office movement (https://www.greenofficemovement.org/) and the Rethinking Economics network (http://www.rethinkeconomics.org/) to mention just a few inspiring examples.
12 With a recognition of epistemological hegemony dividing the soft reform and radical reform responses, and a recognition of ontological hegemony separating the radical reform from the beyond reform space.
13 Thanks to Professor Kevin Anderson for providing this metaphor.
14 For more information on CEMUS, visit www.cemus.uu.se. For a brief overview of CEMUS education, see Stoddard et al. (2012). For a more comprehensive discussion see Hald (2011).
15 To some degree, this already happens at CEMUS as students employed as course coordinators often bring in perspectives and develop pedagogic approaches not usually part of learning at universities.
16 A warm thanks to Professor Vanessa Andreotti for sharing this metaphor with us at the ClimateExistence conference in 2018 (www.climateexistence.se).
17 Those of us that tend to spend time at a university (as students, researchers, faculty or administrators) are quite likely to not belong to this group, which although being a problem, is not an insurmountable one.
18 Seed and Macy et al. (1988) is a good resource for those interested in delving deeper into these questions.
19 Up to 75% per cent of the world's food crops depend at least in part on pollination (FAO, 2016).
20 It is worth noting, however, as a word of caution, that many of these uprisings were met with brutal force and with a backlash that at times created a situation that was worse than the situation was before (just look at many of the Northern African and Arab states today).
21 A few examples being the economics students engaged in the network *Rethinking Economics*, French elite students vowing to not work for companies invested in fossil fuel futures, the Fossil Free and Divestment movements at universities, the *Extinction Rebellion* movement, and various other climate justice initiatives engaging in direct action against fossil fuel infrastructure, children taking their governments to court for destroying their livelihoods with their inaction on climate change. And most recently, the growing movement of school strikes, where children under the age of eighteen all around the world are walking out of school to demand a just and scientifically grounded response to climate change.
22 This is one of the inspirations behind a space in the centre of Uppsala that we have set up: www.kollaboratorietuppsala.se.
23 See e.g. Wright (2017).
24 See e.g. *Reclaiming Our University* movement and manifesto, https://reclaimingouruniversity.wordpress.com/.
25 See e.g. Macy and Johnstone (2012) and Jensen (2006) for further thoughts on this question.
26 Read more at www.climateexistence.se.
27 With institutions such as the *National Touring Theatre* in Sweden, the *Uppsala Art Museum* and the *Museum of Modern Art* in Stockholm.

Acknowledgements

I wish to thank all of the people who took part in the conversation by Lake Balaton for wise and honest reflections that guided me in the initial drafting of this chapter. Thank you Christina Apetrei, Poonam Batra, Keri Facer, Diego Galafassi, Daniel Lang, Heila Lotz-Sisitka, Christoph Mandl, Mark McCaffrey, David Orr,

Jan Sendzimir, Chirapol Sintunawa and Karim Yassin-Goessinger. A big thank you also to Stuart Hill, David Wright, Jacob Rask, Lakin Anderson, Jan van Boeckel, Ingrid Rieser, Keri Facer, Maj Stoddard, Niclas Hällström and Per Johansson for valuable conversations and comments. To all my dear friends and storying wanderers – this piece would not have come about of Norway were it not for our convivial adventures along the mountains, valleys and fjords. A special thanks also to Kevin Anderson whose research and integrity is a continuous source of inspiration. Thanks to David Orr and Dougald Hine for allowing me to draw upon your poetic sensibilities for the title of this chapter. To all my courageous and inspiring colleagues at CEMUS, a heartfelt thanks for your companionship and support in storms, as well as great sailing in the academic waters of Uppsala.

References

Abram, D. (1996). *The spell of the sensuous: Perception and language in a more-than-human world.* Vintage Books.

Anderson, K. (2018a). *Response to the 1.5°C special report.* CEMUS, Uppsala University and SLU. http://www.web.cemus.se/response-ipcc-report/

Anderson, K., & Peters, G. (2016). The trouble with negative emissions. *Science*, 354(6309), 182–183.

Anderson, K., Broderick, J. & Stoddard, I. (2020). A factor of two: how the mitigation plans of 'climate progressive' nations fall far short of Paris-compliant pathways, Climate Policy, DOI: 10.1080/14693062.2020.1728209

Anderson, L. (2017). *Creating spaces for sustainability transformation: Conditions and practices at a transdisciplinary, student-driven center for sustainability in Uppsala, Sweden.* [Master's thesis, Swedish University of Agricultural Sciences]. Epsilon Archive for Student Projects.

Andreotti, V., Stein, S., Ahenakew, C., & Hunt, D. (2015). Mapping interpretations of decolonization in higher education. *Decolonization: Indigeneity, Education & Society*, 4(1), 21–40.

Chancel L., & Piketty, T. (2015). *Carbon and inequality: From Kyoto to Paris: Trends in the global inequality of carbon emissions (1998–2013) and prospects for an equitable adaptation fund.* Paris School of Economics. http://piketty.pse.ens.fr/files/ChancelPiketty2015.pdf

Evernden, N. (1993). *The natural alien: Humankind and the environment.* University of Toronto Press.

Facer, K. (2018). The University as engine for anticipation: Stewardship, modelling, experimentation and critique in public. In R. Poli (Ed.), *Handbook of anticipation.* (pp. 1–20) Springer.

Facer, K. (2020). Convening publics? Co-produced research in the entrepreneurial university. *Philosophy and Theory in Higher Education.*

FAO (2016). *Pollinators vital to our food supply under threat.* Food and Agricultural Organization of the United Nations. http://www.fao.org/news/story/en/item/384726/icode/

Gardiner, S. (2011). *A perfect moral storm: The ethical tragedy of climate change.* Oxford University Press.

Hald, M. (Ed.). (2011). *Transcending boundaries: How CEMUS is changing how we teach, meet and learn.* CEMUS and Uppsala University. http://www.web.cemus.se/wp-content/uploads/2015/05/TranscendingBoundaries.pdf

Hickel, J., & Kallis, G. (2019). Is green growth possible? *New Political Economy*, https://doi.org/10.1080/13563467.2019.1598964

Hill, S. B., Wilson, S., & Watson, K. (2004). Learning ecology: A new approach to learning and transforming ecological consciousness: Experiences from social ecology in Australia. In E. V. O'Sullivan & M. Taylor (Eds.), *Learning toward an ecological consciousness: Selected transformative practices* (pp. 47–64). Palgrave Macmillan.

IPCC (2014). *Climate change 2014: Synthesis report.* Intergovernmental Panel on Climate Change. http://www.ipcc.ch/report/ar5/

IPCC (2018). *Global Warming of 1.5°C.* Intergovernmental Panel on Climate Change. http://ipcc.ch/report/sr15/

Jensen, D. (2006,May 2). Beyond hope. *Orion Magazine.* https://orionmagazine.org/article/beyond-hope/

Kingsnorth, P., & Hine, D. (2009). *The dark mountain manifesto.* The Dark Mountain Project.

Leopold, A. (1949). *A sand county almanac.* Oxford University Press.

Macy, J., & Johnstone, C. (2012). *Active hope: How to face the mess we're in without going crazy.* New World Library.

Masschelein, J. (2010). Educating the gaze: The idea of a poor pedagogy. *Ethics and Education,* 5(1), 43–53.

McLaren, D. (2018). Whose climate and whose ethics? Conceptions of justice in solar geoengineering modelling. *Energy Research & Social Science,* 44, 209–221.

Monbiot, G. (2015,December 12) Grand promises of Paris climate deal undermined by squalid retrenchments. *The Guardian.* https://www.theguardian.com/environment/georgemonbiot/2015/dec/12/paris-climate-deal-governments-fossil-fuels

NOAA (2018). *2018 annual global climate report.* National Oceanic and Atmospheric Administration. https://www.ncdc.noaa.gov/sotc/global/201813

Orr, D. (1991). What is education for? Six myths about the foundations of modern education and six new principles to replace them. *In Context,* 27, 52–57.

Parrique T., Barth J., Briens F., C. Kerschner, Kraus-Polk A., Kuokkanen A., & Spangenberg J. H. (2019). *Decoupling debunked: Evidence and arguments against green growth as a sole strategy for sustainability.* European Environmental Bureau.

Seed, J., Macy, J., Fleming, P., & Naess, A. (1988). *Thinking like a mountain: Towards a council of all beings.* New Society Publishers.

Steffen, W., Rockström J.*et al.* (2018). Trajectories of the Earth system in the Anthropocene. *Proceedings of the National Academy of Sciences of the United States of America.* https://doi.org/10.1073/pnas.1810141115

Stoddard, I. (2017a). The forest knows where you are. *The CEMUS Diaries – stories from past, present and future.* CEMUS and Uppsala University. http://www.web.cemus.se/diaries/#week18

Stoddard, I. (2017b). Copy but don't paste: From Student-led to collaborative action for sustainability in higher education. *LUMAT-B: International Journal on Math, Science and Technology Education,* 2(3), 27–32.

Stoddard, I. Rieser, I., Andersson, S., & Friman E. (2012). Igniting a learning revolution: Student-run education for sustainable development. *Solutions,* 3(5), 34–39.

Wagoner, D. (1996). Lost. In D. Whyte (Ed.), *The heart aroused: Poetry and the preservation of the soul in corporate America* (p. 259). Currency Doubleday.

Wetter, J.K., & Zundel, T. (2018). *The big bad fix: The case against climate engineering.* ETC Group.

Wright, D., Camden-Pratt, C., & Hill, S. B. (Eds.). (2011). *Social ecology: Applying ecological understanding to our lives and our planet.* Hawthorn.

Wright, S. (2017). Can the university be a liveable institution in the Anthropocene? In R. Deem & H. Eggins (Eds.), *Higher education research in the 21st-century series: Vol. 10: The university as a critical institution?*Sense Publishers.

WMO (2019). *WMO provisional statement on the state of the global climate 2019*. World Meteorological Organization. https://library.wmo.int/doc_num.php?explnum_id=10108
WMO (2020). *WMO confirms 2019 as second hottest year on record* [Press Release]. World Meteorological Organization. https://public.wmo.int/en/media/press-release/wmo-confirms-2019-second-hottest-year-record
WWF (2018). *Living planet report 2018*. World Wildlife Foundation. http://wwf.panda.org/knowledge_hub/all_publications/living_planet_report_2018/

5

WILD PEDAGOGIES AND THE PROMISE OF A DIFFERENT EDUCATION

Challenges to change

Bob Jickling and Sean Blenkinsop

So, what is wild pedagogies about?

To begin, we do not want to portray wild pedagogies as a plan, a framework, or a set of guidelines, but rather as a heuristic. We don't introduce this term lightly. The terms *framework* and *heuristic* are different in important ways that go right to the heart of our project.

We see frameworks as providing more concrete visions about how things are, how they should be, or roadmaps for getting to a new place. As such, they assert more control over analysis and can be more prescriptive. But heuristics are typically conceived as agents in the process of discovery. And they even allow challenges to the kinds of control inherent in frameworks. They can act as aids to understanding or even shortcuts into the work itself. Heuristics are provocateurs at the intersection of imagination and praxis. Their aims are more expressive and generative – more attuned to the *wild* reader.

As a heuristic, we see wild pedagogies working as a collection of narratives that begins to do the work that will challenge conventional thinking, and lead willing educators into new territories. Wild pedagogies offer avenues forward for educators determined to try something different, to break free from the hidden authorities within schooling that tend to stop innovations or bend them back towards everyday norms. In the process, we expect – nay encourage – readers and interested practitioners to grow, expand, develop, and quarrel with the concepts outlined here.

Wild pedagogies arose out of a convergence of ideas about wilderness and education in the context of an emerging geological epoch: the Anthropocene. This work is at least in part about reclaiming language and ideas that have been put aside for too long. We will now sketch some of our underlying conceptions about wilderness and education, and some of our pedagogical responses, collectively known as the six touchstones. We will then move to add a new touchstone: the imagination.

On wilderness

We recognise that the idea of wilderness has been roundly and rightly criticised. However, the everyday acceptance of this critique – in some quarters at least – seems to have outstripped its strength. At times, casual dismissals of all wilderness talk as naïve romanticism seem to roll off the tongue with little more weight than an average cliché. That doesn't mean that the original critiques weren't valuable – they surely were. Rather, we may have become too comfortable with them and it is again time to re-think wilderness. There are disruptive anomalies. For those who travel the outback regions of the world, there are still wild places. Wilderness is more than just an idea. And, for others, physical wildness is present much closer to home – even in urban areas.

Inspired by Deleuze and Guattari's (1994) idea that concepts are constantly being created, it seems timely to think again about what wilderness is becoming, or could become. We certainly acknowledge the colonial legacy of wilderness and the leverage this has had in the disenfranchisement of people, families, and cultures the world over. And we stand by this critique. However, wilderness does not necessarily rely on a pristine absence of people for its meaning.

We also recognise that wilderness is sometimes used in a way that reduces its value to a backdrop for human-centred and self-serving ends. This is also a problem, especially in a time of ecological crisis. Again, the conquerable frontier doesn't seem to be essential to the ongoing creation and re-creation of wilderness. Indeed, wild pedagogues have often gone out of their way to see the more-than-human world as a partner and co-teacher (Crex Crex Collective, 2018).

Despite liabilities, wilderness seems to persist as a potentially useful concept. Some sense of this persistence finds support in etymological tracings of the word. For example, *wilderness* can be derived from the Old English *wildoerness*. From this, *wil* can be linked to "wild" or "willed," *doer* to "beast" and, *ness* to "place" or "quality." Putting these together suggests that wilderness can be thought of as a place of wild beasts, or more evocatively, *self-willed land* (Foreman, 2014). When this idea of self-will is juxtaposed with the idea of domestication, a different component of wilderness is given more weight. In this instance *domesticate* is used in the sense of having been brought under control by humans (Livingstone, 1994). We acknowledge that wilderness and domestication should not be thought of as absolute qualities and that wildness occurs in varying degrees. Capacity for self-will, or wildness, hints at concepts like "freedom," "flourishing," "self-determination," and "intrinsic value." For wild pedagogies, it also helps to problematise ideas related to control.

On education

David Orr (2017), like many others, has persistently called for serious educational change. Recently he said:

> Without exaggeration it will come down to whether students come through their formal schooling as more clever vandals of the Earth and of each other or

> as loving, caring, compassionate, and competent healers, restorers, builders, and midwives to a decent, durable, and beautiful future.
>
> *(pp. ix–x)*

What then will it take to nurture healers and restorers of the earth?

While many readers seek to answer Orr's call to make a difference, they participate in systems that are inherently hostile to the task. Enormous forces are at play that, in various combinations, tend to undermine innovations, or bend them back towards the *status quo*. Often unwittingly, these forces tend to control what can happen educationally. Consider the limitations imposed by a global testing culture, an emphasis on particular forms of evaluation, predetermined learning outcomes, an assumed culture of practice, the ubiquitous school-building, and the limited breadth of educational experience encountered – even by willing and enthusiastic practitioners.

Wild pedagogies offer one avenue forward for educators determined to try something different – to break free from the hidden authorities that tend to control schooling. Here we turn to a series of touchstones to be considered as reflective agents and pedagogical aids to change.

On touchstones

Touchstones are concepts and questions that are intended to support and sustain the work of wild pedagogues. They recognise the difficulty in enabling sustained cultural change by providing ideas that can be held, and returned to over and over for guidance, reference, and support. Touchstones are reminders of what wild pedagogues are trying to do, especially when daily demands become overwhelming or when they feel stuck. However, touchstones are meant to be agents of discovery and support, not prescriptions. We suggest reading, responding, and revising them as part of an evolving, situated, and lived practice.

Touchstones are comprised of explanatory text often accompanied by examples drawn from real experiences and experiments. And they attempt to bring the natural world actively into educational conversations – as partner, and co-teacher.

The final element of each touchstone is a series of questions that is designed to prompt readers as they go about practicing and developing wild pedagogies. Some sample questions arising from the most recent iteration of wild pedagogies (Jickling et al., 2018) include:

- How can I invite the natural world to be present as a co-teacher in my practice?
- How did my practice today take risks in moving away from the full control of assumed ends? And how might it continue with that tomorrow?
- What room is there for the unknown, spontaneous, and unexpected to appear and be taken seriously in our educational work?
- How can I make it possible for my students to have encounters with the wild and/or self-willed communities around us?

- In what ways can I notice, and respond to human-centred and colonising perspectives that we encounter?
- How does my practice respond to the existing curriculum and values that are embedded in my workplace? How satisfied am I with my response? Why? What are my criteria for satisfaction?

Maple Ridge Environmental School Project

To better understand the challenges to change, we draw on research from a very radical public school. In 2010 concerned researchers, educators, and other allies met to discuss educational change. These discussions led to the determination that the required change needed to be radical, theory-driven, and well beyond simple tinkering with a very un-green educational system. Formation of an alternative, yet public, school was proposed with the underlying intention of creating an environment where the natural world and its denizens were much more present in the educational process, and where the accompanying research sought to determine how, or indeed whether, the prevailing culture of consumption, anthropocentrism, and alienation from the natural world could be transformed.

While the journey towards the creation and opening of the Maple Ridge Environmental School[1] in 2011 is necessarily a long story, for our present purposes, the school successfully opened with specific aims of having no buildings, conducting all learning outdoors, understanding that the natural world would be part of the teaching faculty, and actively questioning every assumption of the mainstream approach to education.

While this is an audacious project to say the least – and isn't audacity what is required now – it successfully remains a public school within a Canadian school district. As of September 2018, it has 88 students (aged four to twelve), four full-time teachers, two part-time teachers in support roles, three educational assistants, a principal, and a waiting list of almost 100.

Seven years after its inception, three authors – all of whom played significant roles in its development – gathered to discuss the project, in the time-honoured spirit of recorded research dialogues. They considered the influences, challenges, and possibilities of policy that impacted, and continue to impact, the Maple Ridge Environmental School Project (Blenkinsop, Maitland, & MacQuarrie, 2019). With their considered experience they identified four different kinds of policy that arose as themes in their discussions. These policy themes had clear implications for the creation and ongoing maintenance of the school, and they have implications for continued development and success of wild pedagogies.

Educational imagination: A touchstone for change

A closer look at the results of this Maple Ridge policy research suggests that imagination is a pressing and under-considered element in change processes and becomes, in wild pedagogies, a suitable touchstone for ongoing consideration.

The authors of this study note that at least two major policy elements needed negotiation in the process of creating the school. First there are requirements that are explicit, laid out in written form and maintained by the governing bodies involved in public schooling (e.g. school district, various levels of government, unions, etc.). Second, there were implicit policies. These are the assumed expectations that were present though unstated. You could say that they are assumptions buried in the cultures of schools and educational professionals. Explicit policies can be easiest to deal with through negotiation, compromise, and in some instances simple compliance.

Implicit policies are much more complex and need to be teased out and contextualised. In this research, the authors found implicit policy fell into three categories: assumed, tradition and self-limited imagination. The first two overlap, but are separated by a sense of how the policy itself is understood. Assumed policies are positions that are not actually written, but are still taken for granted such that everyone assumes they are required. However, in the case of tradition, while the policy is understood as unstated, it exists as "the way things have always been done." There is a weight of history and inertia connected to tradition.

The third category, self-limited imagination, was a bit of a surprise. In hindsight, it should not have been. This sub-category of implicit policy covers situations, innovative policies, or different operating systems that seem beyond the imagination. This was not a case of the policy being envisioned and then rejected or deemed impossible. This was more about alternative policies not being imaginable at all – about an imaginary limit being reached. When something beyond these imaginary boundaries was offered, the response was often complete blankness, or the muttered "I have never even thought of that."

For us, the idea of a self-limited imagination is striking. And, when not addressed, it stands to thwart far-reaching or radical innovation, and indeed wild pedagogies. We must be aware that the imagination is not broad and flexible as suggested in quotidian understanding. For the rest of this chapter, we explore imagination, its possibilities, and its limitations as the basis of a touchstone for innovators, and wild pedagogies.

Touchstone: The imagination, limits and possibilities

> We believe that ecologically the world has changed and that education can no longer solely focus on preparing students for the tried and true ways of being.
> Expanding our imaginative range is important for creating the communities of the future.

We start from the position that the current modern Western, globalising models of social organisation and economies are deeply flawed and unsustainable. Thus, we will need to become better at imagining, and then enacting, other possibilities. Imagining and enacting are paired: it doesn't do us much good to invent fantastic new worlds if we can't see ourselves creating them, and being in them. To paraphrase Sartre, if we cannot actually see ourselves living in a different way, then the best we can expect is to adapt to the way we currently live. Adapt is a striking word in our ecologically fraught context. As we know, those that cannot adapt will die. The stakes are high.

It is important, however, to be clear that imagination, as explored in the policy discussion, is not an infinite realm in which anything is possible. The imagination is limited by our histories – cultural, experiential, and creative. There are always things, ways of being, and ideas that we cannot imagine – limits to our imaginative capacities. The question then becomes: How do we expand our imaginative range?

In its first year of existence, the faculty and research team at Maple Ridge Environmental School decided that it would be ideal if students were allowed significant unstructured play time in the forest, working on forts. Research (Sobel, 2001) suggested that time with nature, child-centeredness, and constructing forts were important, even necessary parts to building environmental relationships. Yet something odd began to happen within a month of having at least an hour per day in the "forts village."

Systems of currency and governance began to develop as the children's imaginations created a community. Within two months the fort village had become akin to an authoritarian police state with one older boy assuming the role of leader supported by a posse of henchmen and bodyguards. The buildings had become jails, casinos, shopping complexes. Resources such as sticks, bits of rope, and properties were being hoarded by particular members of the leadership group. The natural world was now a mere resource for individual enrichment, and particular areas of the village were denuded of life. It is important to note that not all the student voices were in line with the macro-narrative at play, but those outside tended to be younger, more marginalised players.

As this community structure started to make itself manifest, teachers began to engage more actively, governing committees and councils were created and the shape of the village changed – a little. But, interestingly, there was only minor change. The teachers themselves were having a hard time imagining what a different kind of community might look like. Like the children, they were having to use the governance and community-making tools that were available to them – those of the culture in which they were immersed. Teachers and students alike were imaginatively limited to the cultural and experiential realities of their lives, and these did not seem to include viable images of more equitable, eco-centric villages in action. Intriguingly, even after spending enormous amounts of time in the natural world, which does not tend to govern itself in the same ways as those emerging in the fort village, the quiet voices of both the marginal children and the natural world were being ignored.

For Russian psychologist Lev Vygotsky (1978), human development is a socio-cultural event and, for our discussion, this has implications for imagination. We are born into a culture that offers psychological and sense-making tools that help us to understand, and then position ourselves within, our world. Tools such as language, story, and even humour are incorporated into the selves that we become. These tools assist us in understanding the world and help us belong to the cultures and communities in which we find ourselves. It is important to note that not only do we gather these sense-making tools and begin to use them, but that in the process we are also being shaped by the tools themselves. The languages we learn to speak and the foundational stories we are told shape who we are in the world.

Kieran Egan (1997), a philosopher of education, interprets Vygotsky in his own work in interesting ways. For Egan, it is the imagination that plays the intermediary role between the individual and their culture. It is the imagination that reaches out into the culture and draws in these cognitive tools. Thus, as the individual uses these tools to make sense of the world, they are also being culturally shaped – and so too is their imagination. We think that in this interplay between individual and culture, the imagination plays a significant role, but it is also being limited in culturally specific ways. It is here, in the expansion of imaginative possibility, that important wild pedagogical work can be done.

It is important to note that neither Vygotsky nor Egan suggest tools that are not cultural in origin. This can leave environmental theorists wondering about a whole range of nature-based sense-making tools that might exist. Things such as binocular vision, biophilia, or upright stance all clearly influence, affect, and shape how we make sense of the world. We also live in different landscapes, ecosystems, and environments. Might these places have different tools to offer, and thus shape us in different ways? And how do our experiences in these places limit our imaginative range? The following unpublished reflection, by co-author Blenkinsop, was written to explore these questions.

> Deep in the boreal forest of Northern Ontario, under the continuous canopy of black spruce, and resting softly upon the sodden mossy carpet that appears to hold the entire forest up, there lies the creeping snowberry. A low-lying vine-like plant, it spreads itself across the forest floor and every once in a while produces tasty white, ant egg-shaped fruit. And it is the creeping snowberry that reminded me of the lesson of the limits of my imagination. As I explored for snowberries, I came to think that there was no rhyme nor reason for their production. There would be only one fruit-producing plant in an entire patch. I couldn't understand, knowing that the snowberry can reproduce asexually, why any single plant would expend the energy to produce fruit – especially when that expenditure puts it at such a disadvantage in relation to its close competitors, other snowberry plants.

Part of an explanation for Blenkinsop's failure to understand the snowberry might rest within the imaginative limits placed upon him by foundational cultural narratives. Whether we like it or not, many of us are incubated in derivatives of Western Judeo-Christian stories (White, 1967). Of interest here is the notion that the world is made up of clear and definable individuals. We hear this in the creeping snowberry discussion where the author posits visible individuality (i.e. snowberries are determined to be separate individuals based on their above ground separation) in his musings.

Interestingly, recent research on the eastern creeping snowberry suggests that any above-ground analysis risks misunderstanding subterranean interrelationships. It turns out that creeping snowberry should be thought of as a community that raises fruit. For example, resources are transferred underground to support the seemingly individual producer. Ecologically speaking, the possibility of having the community's genes disbursed beyond the normal reach of the roots is worth the effort. This is because the seemingly separate plants in any patch are in fact interconnected and related to each other and thus having a sexually produced fruit eaten and then dispersed by a passing animal is a benefit to the entire community. Those raised within an individualistic ethos can lack the capacity to easily imagine such a community-based orientation. It seems that individuality is fundamental to our noun-based languages, to our economic and political systems, to the epistemology upon which public schooling tends to be based, and to many of our foundational stories.

Indigenous–Greek scholar Thomas King illustrates these points in another way. In his Massey lecture series, *The Truth About Stories* (2008), he suggests that all we are is stories. In ways that seems to resonate with Vygotsky, King shows us how we become the stories that we are immersed in, our culture chooses to tell us, we tell ourselves, and we are told to us by marketers, politicians, family, and teachers.

King illustrates his discussion by placing the Genesis creation story side-by-side with an Indigenous creation story. The contrast reveals possibilities for enacting radically different ways of being in the world. The Genesis creation story presents an omnipotent, all-knowing, male God who makes all the decisions, whereas the Indigenous story presents a woman leader in conversation and negotiation with already existing animals and birds. For King, these foundational stories have deep implications for the cultures they sustain. Each foundation offers possibilities. Each assists individuals in making sense of the world, but neither is opening a panoply of what it might mean to be human. There are limitations to each and as a result, those who are shaped by these stories and languages are limited as well in, amongst other things, their imaginative capacities.

For rebel educators and wild pedagogues this challenge of expanding one's own imagination and those of their students has implications for how we imagine and then create our own lives as teachers. Blenkinsop (2012) suggests a Foucauldian stance of "hyperactive pessimism." Here the challenge is to increase one's vigilance and self-reflexivity in everything related to practice. At the same time, knowing how ineffective modern Western education has been in engaging in environmental matters, one is likely to misstep along the way. One pitfall that endangers us all is our "normal," "common sense" intuitions, languages and stories. When we realise that these tendencies have grown out of our own histories, we can understand that many of our first impulses are likely to be ones that have been shaped by the very *status quo* that we seek to challenge. The point is that we have to watch everything we do, and we should expect to find in our practices things that we would rather not do.

We should also expect that while our imaginative capacity will always be limited, there are ways to expand our reach. This will require a humble orientation; a willingness to change; an active gathering of other ideas for how to be in the world both within one's cultural reality and beyond; a constant expanding of the tools, both cultural and natural, cognitive and physical, that are being made available; a careful consideration of the stories, metaphors, and languages one is using; and a thoughtful engagement in an ever-widening range of experiences. The last consideration is proposed not to create students who run thoughtlessly through hundreds of new adventures, but because the imagination relies on the cultural tools that it draws into the learner and the "stuff" with which it has to work. This includes ideas, concepts, experiences, encounters, etc., and it is up to wild pedagogues to carefully consider their learners and offer that which might help to expand their imaginative potential.

With that we conclude with some questions the wild pedagogue might want to consider:

- What did I do with my practice today that pushed outside the students' and my imagination?
- What new "stuff," experiences, and stories did I add to the mix? And how are students taking up, working with, and being changed by what is added?
- How well did I notice my proclivity to not do the seemingly unusual and then make a considered attempt at it anyway?
- What cognitive, physical–cultural and natural tools are my students working with right now? And what new ones might I try introducing?
- What are the edges of my experience that might limit how far I can imagine a different kind of education for my students? What are the limits of my own imagination?
- What are the sources of inspiration (e.g. experiential, trans-cultural, literary, etc.) that I am seeking to support and enhance my process of pedagogical change and development?
- In what ways do I have a better sense of the edges of the imagination that exist in my school, community and larger culture?

Note

1 See https://es.sd42.ca/.

References

Blenkinsop, S. (2012). Four slogans for cultural change: An evolving place-based, imaginative, and ecological learning experience. *Journal of Moral Education*, 41(3), 353–368.

Blenkinsop, S., & Morse, M. (2017). Saying yes to life: The search for the rebel teacher. In R. Jickling & S. Sterling (Eds.), *Post-sustainability and environmental education: Remaking education for the future* (pp. 49–61). Palgrave Macmillan.

Blenkinsop, S., Maitland, C., & MacQuarrie, J. (2019). In search of policy that supports educational innovation: Perspective of a place- and community-based elementary school. *Policy Futures*, 17(4), 489–502.

Crex Crex Collective: Hebrides, I., Affifi, R., Blenkinsop, S., Gelter, H., Gilbert, D., Gilbert, J., Irwin, R., Jensen, A., Jickling, B., Knowlton Cockett, P., Morse, M., Sitka-Sage, M., Sterling, S., Timmerman, N., & Welz, A. (2018). Six touchstones for wild pedagogies in practice. In B. Jickling, S. Blenkinsop, N. Timmerman, & M. D. Sitka-Sage (Eds.), *Wild pedagogies: Touchstones for re-negotiating education and the environment in the Anthropocene*. Palgrave Macmillan.

Deleuze, G., & F. Guattari. (1994). *What is philosophy?* (G. Burchell & H. Tomlinson, Trans.). Verso. Columbia University Press

Egan, K. (1997). *Educated mind: How cognitive tools shape our understanding*. University of Chicago Press.

Foreman, D. (2014). *The great conservation divide: Conservation vs. resourcism on America's public lands*. Ravens Eye Press LLC.

Jickling, B., Blenkinsop, S., Timmerman, N., & Sitka-Sage, M. (2018). *Wild pedagogies: Touchstones for re-negotiating education and the environment in the Anthropocene*. Palgrave-MacMillan.

King, T. (2008). *The truth about stories: A native narrative*. University of Minnesota Press.

Livingstone, J. A. (1994). *Rogue primate: An exploration of human domestication*. Key Porter Books.

Orr, D. (2017). Foreword. In B. Jickling & S. Sterling (Eds.), *Post-sustainability and environmental education: Remaking education for the future* (pp. vii–xi). Palgrave Macmillan.

Sobel, D. (2001). *Children's special places: Exploring the role of forts, dens, and bush houses in middle childhood*. Wayne State University Press.

Vygotsky, L. (1978). *Mind in society: The development of higher psychological processes*. Harvard University Press.

White, L.Jr. (1967). The historical roots of our ecologic crisis. *Science*, 155, 1203–1207.

PART 2

Transforming practice

6

COMMUNITY EDUCATION AND PARTNERSHIPS FOR SUSTAINABLE DEVELOPMENT

A way forward for Indigenous Asia

Subarna Sivapalan and Ganakumaran Subramaniam

Introduction

Projections by the World Bank in 2019 indicate that the impacts of climate change could push in excess of 100 million people into poverty by 2030. Indigenous people, who make up five per cent (370 million) of the total global population, are at great risk, as they also make up one sixth of the world's most socially and economically challenged, and a third of the world's 900 million destitute poor (IndigenousPeoples' Major Group for Sustainable Development, 2019).

Climate change poses a severe threat to Indigenous community development efforts around the globe. These communities safeguard approximately 80 per cent of biodiversity, and sustainably oversee as much as 50 per cent of the world's lands. Ironically, their rights to ownership of these lands are merely 10 per cent (IndigenousPeoples' Major Group for Sustainable Development, 2019). This 2019 report has further stressed that although most Indigenous people live in rural areas, national governmental economic growth plans and interventions often overlook territories that are predominantly Indigenous. The lack of commitment towards finding sustainable solutions to this conundrum will have increasing adverse impacts on the survival of these communities.

The Indigenous agenda has received some attention at the global scale. International frameworks central to this agenda include the 1957 International Labour Organization's Indigenous and Tribal Populations' Convention, the 1989 International Labour Organization's Indigenous and Tribal Peoples Convention, the 2007 United Nations Declaration of the Rights of Indigenous Peoples, the 2014 World Conference of Indigenous Peoples, and the 2015 Paris Agreement. The United Nations Permanent Forum on Indigenous Issues, the key advocate for Indigenous voices globally, has also declared that the sustainable development goals (SDGs) and their associated indicators are relevant to the rights of Indigenous peoples. The 2017 IndigenousPeoples' Major Group for Sustainable Development High Level

Political Forum Report noted six references to Indigenous populations in its "2030 Agenda Resolutions," namely empowerment, engagement and education of Indigenous peoples within paragraphs 23, 25 and 52 of the preamble, target 2.3 of goal 2 ending hunger through sustainable agriculture, target 4.5 of goal 4 ensuring access to education for Indigenous peoples, and paragraph 79 of the resolution's Indigenous peoples' participation in follow-up and review (United Nations, 2017a). Other key priorities for Indigenous peoples, of which the formulation of the "2030 Agenda" is considered as a mechanism to uplift the community, are land rights, the elimination of poverty and hunger, social security, health and education, environmental sustainability, promotion of inclusive and peaceful societies, reduction of inequalities and overcoming discrimination.

Ironically, although Indigenous issues feature within the "2030 Agenda," the SDGs present both irrefutable benefits as well as challenges for Indigenous communities. Many of these challenges relate to Indigenous land and territories and to recognition of the communities':

- collective rights in areas such as health, education, culture and ways of living
- self-determination to define their economic, political, social and cultural development
- holistic development
- right to exercise the principles of free, prior and informed consent
- culture-sensitivity in areas such as health and education (United Nations, 2017b).

Two years on, the IndigenousPeoples' Major Group for Sustainable Development (2019) noted the following:

> From the perspective of Indigenous peoples, inclusion and empowerment entail legal recognition of their distinct identities; security of tenure of their lands, territories and resources; peace in their territory and enjoyment of the right to self-governance including their customs, traditions, cultures, and livelihoods linked to sustainable resource management practices.
>
> *(p.2)*

These observations make clear the need to accentuate systemic and sustainable interventions for the Indigenous community at global, regional and national levels, focusing mostly on participatory approaches that aim to better understand and support the needs of Indigenous communities. Such strategies are particularly important within the Asian context, where the voices of Indigenous peoples have been largely overlooked and are underrepresented in global sustainability dialogues.

Indigenous Asia at a sustainability crossroads

The Asia and Pacific Indigenous population make up approximately 70 per cent (over 260 million people) of the world's total Indigenous peoples (World Bank,

2018); and they are also the most marginalised (Food and Agriculture Organization, 2018). The authors of this FAO report state: "lack of respect to their basic human rights, cultures, spirituality and traditions, and the encroachment on their traditional lands and natural resources increase their vulnerability by undermining livelihoods, shelters and identity" (p.1). This is cause for concern, especially since Indigenous communities in these regions mostly live below the poverty threshold, are deprived of education, lack an understanding of their rights and are not adequately represented in decision-making processes.

Most Indigenous peoples in Asia live in Japan, Taiwan, China, Philippines, Indonesia, Timor Leste, Brunei, Malaysia, Thailand, Cambodia, Lao PDR, Vietnam, Myanmar, Bhutan, Nepal, Bangladesh, Sri Lanka, India and Pakistan; they represent approximately 2000 distinctive civilisations and languages (Errico, 2017, see Table 6.1). They are often referred to as tribal, highland, native, hill, aboriginal or ethnic minorities (Asia Indigenous Peoples Pact, 2014).

Although the 2007 United Nations Declaration on the Rights of Indigenous Peoples has been adopted in almost all countries in Asia, and there has been growing participation from the Indigenous peoples of Asia in United Nations agencies and consultation processes involving Indigenous issues (Asia Indigenous Peoples Pact, 2014), the Indigenous people of Asia, still have limited visibility within the Association of South East Asian Nations (ASEAN) Human Rights Declaration. Further, Asian Indigenous communities continue to grapple with issues of basic human needs and rights, such as lack of access to quality education, healthcare, electricity and clean water, legal recognition, the right to lands, climate change and sustainable development (Asia Indigenous Peoples Pact, 2014).

It is now acknowledged that the region will not be able to accomplish its promised SDG targets by the year 2030 at the present rates of development in relation to SDG17 partnerships for the goals (United Nations, 2019). This will be particularly challenging given the consequences of climate change for the 200 million plus Indigenous people in the Asia-Pacific region, as noted in the Asia Indigenous Peoples Pact (2014). This report highlights the Indigenous situation as follows:

> Despite having one of the smallest carbon footprints, Indigenous peoples bear the brunt and burden of the impacts of climate change and the effects of flawed solutions being implemented to mitigate it. Their traditional lifestyle and customary resource management systems provide for a prudent and sustainable use of resources, in which they take and replenish only what they need. However, with the degradation and loss of their lands and environment due to wanton resource extraction, their vulnerability to extreme weather conditions has increased; at the same time these have reduced their capacity to cope and adapt to climate change.
>
> *(p.12)*

Within the Indigenous landscape of Asia are the Orang Asli of Peninsular Malaysia, who together with the Orang Asal, make up 12 per cent of the total national

TABLE 6.1 Indigenous Peoples in Asia – Description, ethnic grouping and percentage of total national population

Country	*Description*	*Number of ethnic groups*	*Percentage of total national population (%)*
Bangladesh	Tribal Peoples, Pahari, Jumma, Adivasi, Ethnic Groups and Minorities	45	1.2 to 2.5
Cambodia	Indigenous Minorities	19–21	0.9 to 1.4
India	Scheduled Tribes, Adivasi	622–635	8.3
Indonesia	Masyarakat Adat	More than 700	20–29
Lao PDR	Ethnic Minorities	Around 200, with 49 officially recognised ethnic minorities	35–70
Malaysia	Natives, Orang Asli, Orang Asal	97	12
Myanmar	Ethnic Minorities	135	30–40
Nepal	Indigenous Nationalities, Adivasi, Janajati	More than 80, with 59 recognised Indigenous nationalities	37.1
Philippines	Indigenous Peoples, Indigenous Cultural Communities	110 officially recognised Indigenous peoples	10–15
Thailand	Ethnic Minorities, Hill Tribes, Hill/Mountain People	More than 50, with 10 officially recognised hill tribes	1.5
Vietnam	Ethnic Minorities	More than 90, with 43 officially recognised ethnic minorities	13.8

Source: Errico, 2017, p.15

population (Errico, 2017). The term "Asli" denotes their status as the original people of the country. The 2010 Orang Asli census places their total population at 178,197, within 18 ethnic sub-groups under the categories of Proto-Malay (42.3%), Senoi (54.9%) and Negrito (2.8%). There are 853 officially acknowledged Orang Asli villages in Peninsular Malaysia, comprising 18 ethnic sub-groups (Department of Orang Asli Development, 2016, see Table 6.2).

The highest population of the Orang Asli in Malaysia is found in the states of Pahang and Perak. 37.2 per cent of Orang Asli communities live in interior and forest areas, whereas 61.4 per cent and 1.4 per cent of the community live in forest-fringe or rural areas, and within or close to urban areas respectively (Centre for Orang Asli Concerns, 2012). In 2016, there were 178,197 Orang Asli living in Peninsular Malaysia (see Table 6.3; Department of Orang Asli Development, 2016).

TABLE 6.2 Orang Asli ethnic sub-groups by category

Senoi	*Proto Malay*	*Negrito*
Semai	Temuan	Kensui
Temiar	Semelai	Kintak
Jahut	Jakun	Lanoh
Che Wong	Kanaq	Jahai
Mahmeri	Orang Kuala	Mendriq
Semoq Beri	Seletar	Bateq

Source: Department of Orang Asli Development, 2016, p.11

TABLE 6.3 Orang Asli Population Living in Peninsular Malaysia by State in 2016

State	*Category*			*Total by State*
	Senoi	*Proto Malay*	*Negrito*	
Johor	55	13083	1	13139
Kedah	19	0	251	270
Kelantan	12047	29	1381	13457
Melaka	28	1486	1	1515
Negeri Sembilan	96	10435	0	10531
Pahang	29439	37142	925	67506
Perak	50281	605	2413	53299
Selangor	5073	12511	3	17587
Terengganu	818	41	34	893
Total Orang Asli Population	**97856**	**75332**	**5009**	**178197**

Source: Department of Orang Asli Development, 2016, p.12

The Aboriginal Peoples Act 1954, Act 134, was established "to provide for the protection, wellbeing and advancement of the aboriginal people of West Malaysia" (Commissioner of Law Revision, Malaysia, 2006, p.5). Since 1957, the year the nation achieved independence, many Orang Asli-focused socio-economic reforms and development programmes have been implemented to improve the wellbeing of the community, primarily to help them assimilate within mainstream society. Most of these reform and development programmes have been focused on agriculture, land and skill development (Department of Orang Asli Development, 2011); Abdullah (2011) considers that they have improved the socio-economic status of the Orang Asli to some extent.

Nevertheless, some of these developments have come at a high environmental, social and economic cost for the Orang Asli. As many still live off the land, encroachment into their traditional land has not only left them with depleted resources, but has also made them more vulnerable to the impacts of environmental degradation, deforestation and climate change. Recognising this, greater accountability has been sought from the government (Chow, 2019), but although this is important, there is also a critical need for stronger participation and partnership from all stakeholders – the government, corporate sector, NGOs and the Orang Asli community itself – for effective change to take place. The development of appropriate multi-stakeholder mechanisms that could support the promotion of sustained participation from the Indigenous community is urgently needed.

Overall, there is a critical need for political will and social and economic interventions to be undertaken in ways that do not disrupt the Orang Asli's unique contribution to ecosystem sustainability. Collaborative dialogue and collective decision-making, facilitated via a community education and partnerships for sustainability approach, is the approach we are proposing to address the need for greater multi-stakeholder participation in environmentally, socially and economically uplifting the Indigenous of Asia so that they are not left behind. This is based on our experiences with our "social responsibility programme" with an Orang Asli community in the state of Perak, in northern Peninsular Malaysia.

Community, community education and partnerships for sustainability

There have been many attempts to define "community"; however, most fail to capture the complexities involved. Lindemann (1921; cited in Hanchor & Olumati, 2012) defined community as "any process of social interaction which promotes greater intensive or more extensive attitude which improves co-operation, collaboration and unification" (p. 59). Bola and Bello (1987; also cited in Hanchor & Olumati, 2012) depicts communities as "a territorially bound social system within which people live in harmony, love, intimacy, and share common social, economic and cultural characteristics" (p.59). Because these adequately describe the context within which the Orang Asli community of Malaysia exist, we will use them as the basis of our understanding of the notion of community.

In identifying the aim of community education, Minzey and le Tarte (1972; cited in Akande, 2007) emphasise inculcating and improving community positivity and community living to enable members of the community to collectively identify problems and develop solutions for these problems. Fletcher (1980; cited in Hanchor & Olumati, 2012), describes community education as "a process of commitment to the education and leisure of all ages through local participation in setting priorities sharing resources and the study of circumstance" (p.61), thereby providing pathways to better understand the community's values, development and culture:

> In the light of community education, emphasis is not only on the formal schooling but the informal which takes place at home and other social institution. Personal growth occurs through a series of learning processes, leading to the development of certain capacities, physical, intellectual and moral, which enables the individuals to function as productive and effective members of the society. The informal aspect of education is concerned with training and skill acquisition which is relevant to adults and youths. The central focus of this form of education is in the area of job and skill orientation, political and cultural participation, social and economic responsibilities; acquisition and exhibition of spiritual and moral values.
>
> *(Fletcher as cited in Hanchor & Olumati, 2012, p.61)*

Community education is invariably a form of non-formal education. It is conducted beyond the formal school system to facilitate specific types of learning experiences within community sub-groups, enabling them to engage in social activities, gain employment, elevate income levels and improve their quality of life (Akande, 2007). Ezimah (2004) similarly recognises the aims of community education as improving awareness, propagating understanding and offering the required competences to socially, economically, politically and culturally develop the community. Because such learning has the capacity to advance global, community and individual development, it can help the Indigenous communities of Asia achieve the targets of "Agenda 2030."

The UNESCO Institute for Lifelong Learning strongly advocates the need for community learning and community-based non-formal education to advance sustainable development:

> The 2030 Agenda encompasses all aspects of our lives, which implies that learning, if it is to contribute fully to this agenda, must be seen as both lifelong and life-wide. The relevance of community-based non-formal education and informal learning for children, young people and adults, especially those not in education or from marginalized or disadvantaged parts of society, must be recognised and fostered in every country of the world if the 17 Sustainable Development Goals are to be met.
>
> *(UNESCO Institute for Lifelong Learning, Policy Brief 8, 2017, p.1)*

This policy brief also highlights the benefits of community-based learning for sustainable development within the context of Indigenous communities, stating "A community approach to lifelong learning for sustainable development helps people to re-identify, re-evaluate and further develop local and Indigenous knowledge, based on still-relevant but frequently neglected traditional wisdom, which community based learning can help reclaim" (p.1). Partnerships are key to effective community education for sustainable development, particularly when the scale of the proposed solutions requires multi-stakeholder interventions.

The United Nations General Assembly Resolution A/RES/70/224 (2015), *Towards global partnerships: a principle-based approach to enhanced cooperation between the United Nations and all relevant partners* defines partnership as "voluntary and collaborative relationships between various parties, both state and non-state, in which all participants agree to work together to achieve a common purpose or undertake a specific task and to share risks and responsibilities, resources and benefits" (p.4). Multi-stakeholder partnerships, according to the Global Knowledge Partnership (2003), "pursue a shared vision, maintain a presumption in favour of joint problem-solving, promote a work ethos that exploits mutual self-interest, and adds value beyond that achievable by the principal alternatives" (p.8).

The notions of community, community education and partnerships discussed above collectively form the theoretical basis of the work documented here. The challenges faced in implementing community education and partnerships for sustainability within the Orang Asli of Peninsula Malaysia, and insights into ways in which these challenges can be collectively addressed through Indigenous community-academia-industry-local government-NGO partnerships, are discussed.

Community education and partnerships for sustainability: The Project Asli experience

Project Asli took place in an Indigenous community situated in the district of Batang Padang, Tapah, in the state of Perak. The project was run in partnership with the community, the university at which the lead author is presently based, three industry partners, an environmental NGO, and the local government, represented by officers from the district's Orang Asli Development Department.

At the time of conducting the project, the Indigenous community at the project site comprised approximately 100 individuals, within 13 families (including a mix of male and female senior citizens, adults, youth, teenagers and younger children). The oldest community member was in his late 60s, and the youngest was 8 months old. Most community members were Christian. The males obtained their daily wages from heavy labour on vegetable farms close to the village. To supplement this variable source of income, female community members occasionally sell handicraft and honey harvested from the forest. Major challenges over the past 40 years have been energy poverty, waste management and access to clean water and sanitation.

We became involved with the community in 2016, when a corporate social investment (CSI) project to install a mini solar farm at the village was conducted by the academic partner in collaboration with an industry partner. This was designed to provide the community with electricity to provide light, and heat for cooking, enabling the Orang Asli to get their children ready for school in the morning, and for the children to do their homework at night. Lighting within the shared communal toilets helped them to avoid attacks from the snakes and wild animals that came into the village from the adjacent jungle. The industry partner provided the financial support for the solar installation, and the academia partner provided training on solar farm maintenance and operation for the Indigenous youth. This nurtured a sense of ownership and avoided dependence on partner organisations.

From 2017 to 2018, community education for sustainability projects within the village focused on improving waste management (creating handicrafts from recycled materials and used t-shirts), sanitation (refurbishment of toilets and the shower booth, sponsored fully by an industry partner), access to clean water (installation of a water filtration system sponsored fully by a partner organisation), and food security (the cultivation of vegetables for community consumption via hydroponics). Basic entrepreneurship skills were taught to the youth and women to help make the community more self-sufficient, both environmentally and economically. Community workshops were provided on the importance of waste segregation and recycling (the community practiced open burning as there were no waste collection services), rural sanitation, and pollution management (land and water), to raise awareness of their impacts on their village and their wellbeing. These workshops also provided an avenue for the needs of the Orang Asli to be heard by local authorities. The university partner monitored the progress of the various projects.

In 2019, ethics approval was obtained from the University of Nottingham's School of Education to conduct to record the perspectives of the Indigenous community and partners on the impact and challenges of this initiative. This also provided an opportunity for the team to reflect on the effectiveness of the programme. Data were gathered from semi-structured interviews from project partners and community members. An interview guide was developed to ensure the issues discussed during the interviews were focused on both the outcomes and challenges of the initiative. Other issues raised were also recorded.

Three stakeholder partners representing industry, NGO and the academic team, agreed to provide feedback for the impact assessment. Interviews with partners were conducted off site (in English). Ten community members were also interviewed (face-to-face) within the village: the Indigenous youth leader and nine other Indigenous community members (aged 16–32), including the Tok Batin (village head) and his wife.

In addition to the ethics approval, consent to conduct the interviews within the village was obtained from the Tok Batin. Interviews with the community members were conducted using an informal talking circle approach facilitated by the lead author, who had conducted projects within this community in the past. The necessary measures were undertaken to ensure researcher reflexivity and to avoid bias. The talking circle lasted 52 minutes, and it was audio recorded with permission from the participants.

Our findings suggested that the community education and partnerships initiative brought to the surface not only challenging environment-related issues, but also critical social and economic problems that required multi-faceted solutions. The community is grappling with fundamental environmental issues, some of which are beyond their control (such as the open burning of waste because of the lack of a waste collection system for the village).

Although the community understood the dangers of unsustainable development to the environment, and how this could potentially affect them, they were yet to fully appreciate how recycling could benefit their wellbeing. Because they live off grid, candles, flashlights and diesel powered generators had become a core feature of the community. To compensate for the lack of electricity, the community depends on natural sunlight from 8am to 6pm, candles from 4am to 8am, and generators from 6pm to 9pm. One house had caught on fire from a candle. With the installation of the solar panels, each household was able to save between RM4.90 to RM5 daily, the same as the amount spent on diesel fuel.

Whereas male community members work mostly in heavy labour jobs, the women who work outside of the home do so as vegetable packers and restaurant helpers. Both worked for daily wages, so missing a day of work means no income for that day. Most families receive their income from a single source (from the husbands). Most wives work in the home, particularly looking after the children, and do not receive an income. Older children often work to supplement the household income, and to do this may drop out of school as early as 13 to 15 years old. Children who complete their secondary education then seek employment. In the history of the community, only two youths have pursued pre-university studies (Form 6), and none have gone to college or university (due to the cost). Nevertheless, the community believes that education is the only way for them to break free from the poverty cycle and lead a more comfortable life. Government subsidies for high school education for Indigenous children have not solved this problem because of the inability of the families to maintain a sustainable income.

The main social problem is the migration of the youth to the cities, where they can secure better paid jobs as factory workers, office cleaners, security guards and shop assistants. But by doing this they risk losing sight of the importance of their Indigenous roots, values and heritage. Because these Indigenous values are sacred, the migration of their youth to the cities causes them much grief. In addition, continuing to study post primary level can be challenging for Indigenous students as they experience a sense of otherness as they interact with non-Indigenous students.

The community members and partners felt that the community education and partnerships initiative helped them understand how sustainable interventions could provide income and benefit the community in the following ways:

- reducing the community's financial dependency on the academia-industry-NGO-local government partners, by educating the community on best practices to independently source solutions for obtaining alternative income

- providing alternatives to formal education via environmental preservation training, which could then act as an alternative source of income for the community
- providing sustainable measures to conserve and preserve Indigenous land from environmental degradation using traditional Indigenous knowledge and values, without having to depend solely on the local council
- providing ways in which Indigenous communities could practise environmental sustainability via Indigenous entrepreneurship initiatives, such as community recycling
- providing sustainable community partnership action plans where the community partner drives the initiative instead of the academia-industry-NGO-local government partners taking the lead, as is presently the case
- recognising Indigenous community leaders as sustainability champions
- rewarding Indigenous community leaders for their efforts in elevating the socio-economic status of the communities via sustainability-focused efforts
- empowering Indigenous communities to take ownership and lead community education for sustainability initiatives, with partners providing support in the form of environmental, social and economic expertise whenever necessary

Conclusions

The global community has an important role to play in implementing "Agenda 2030". Community education and partnerships for sustainability provide a suitable way to achieve this. Here we have documented some of the potential benefits of employing community education and partnerships for sustainability to contribute towards the targets of the Sustainable Development Goals. We recognise the challenges to Indigenous communities in achieving these targets and have proposed ways in which community education and community-academia-industry-NGO-government partnerships can pave the way for Indigenous communities, particularly for those in Asia, to address these challenges. International and local legislative and structural policies and processes need to be further improved to ensure that Indigenous communities are not left behind in our quest to achieve global goals, particularly through partnership personnel training to ensure that interventions and communications are socially and culturally sensitive and non-intrusive or judgemental to Indigenous people.

Acknowledgements

This work is part of a larger UNESCO global project focused on the Indigenous community and education for sustainable development aimed at reorienting education and training systems to improve the lives of Indigenous youth. The project is coordinated by the UNESCO Chair in Reorienting Education towards Sustainability, York University, Toronto, Canada. This work was partially funded by the 2019 cycle of the Fundamental Research National Grant, and the Conservation Partnership Scheme grant, by WWF Malaysia.

References

Abdullah, R. (2011). Poverty among the Indigenous people (Orang Asli): Cases in Terengganu. In I. C. Omar (Ed.), *Issues and Strategies in Poverty Eradication*. Universiti Malaysia Kelantan.

Akande, J. O. (2007). The practice of community education in Nigeria. *Educational Research and Review*, 2(10), 264–270.

Asia IndigenousPeople Pact. (2014). *Overview of the state of Indigenous peoples in Asia*. www.aippnet.org

Centre for Orang Asli Concerns. (2012). *The Orang Asli of Peninsular Malaysia*. https://www.coac.org.my/main.php?section=about&page=about_index

Chow, M. D. (2019April 21). Stop hurting environment, pressure groups tells policymakers. *Free Malaysia Today*. https://www.freemalaysiatoday.com/category/nation/2019/04/21/stop-hurting-environment-pressure-group-tells-policymakers/

Commissioner of Law Revision, Malaysia. (2006). *Laws of Malaysia Reprint, Act 134, Aboriginal Peoples Act 1954*. https://www.jkptg.gov.my/images/pdf/perundangan-tanah/Act_134-Oboriginal_Peoples_Act.pdf

Department of Orang Asli Development (JAKOA). (2011). *Orang Asli Strategic Development Plan 2011–2015*. Department of Orang Asli Development, Malaysia.

Department of Orang Asli Development (JAKOA). (2016). *Strategic Plan 2016–2020*. http://www.jakoa.gov.my/wp-content/uploads/2019/12/pelan-strategik-JAKOA.pdf

Errico, S. (2017). *The rights of Indigenous peoples in Asia: A human rights-based overview of national legal and policy frameworks against the backdrop of country strategies for development and poverty reduction*. International Labour Organization. https://www.ilo.org/wcmsp5/groups/public/—dgreports/—gender/documents/publication/wcms_545487.pdf

Ezimah, M. O. A. (2004). *Knowing adult education: Its nature, scope and processes*. Springfield Publishers.

Fletcher, C. (1980). The theory of community education and its relation to adult education. In J. L. Thompson (Ed.), *Adult Education for a Change*. Hutchinson.

Food and Agriculture Organization. (2018). *Indigenous peoples in the Asia-Pacific region*. http://www.fao.org/3/CA2045EN/ca2045en.pdf

Global Knowledge Partnership. (2003). *Multi-stakeholder partnerships issue paper*. https://www.odi.org/sites/odi.org.uk/files/odi-assets/publications-opinion-files/2117.pdf

Hanchor, M. E., & Olumati, E. S. (2012). Enhancing community development through community education. *Journal of Education and Practice*, 3(14), 59–63.

IndigenousPeoples' Major Group for Sustainable Development. (2019). *Inclusion, equality and empowerment to achieve sustainable development: Realities of Indigenous people. United Nations High Level Political Forum Report*. https://www.indigenouspeoples-sdg.org/index.php/english/all-resources/ipmg-position-papers-and-publications/ipmg-reports/global-reports/124-inclusion-equality-and-empowerment-to-achieve-sustainable-development-realities-of-indigenous-peoples/file

Lindemann, E. C. (1921). *The community: An introduction to the study of community leadership and organization*. Associated Press.

United Nations. (2015). *General Assembly Resolution A/RES/70/224 (2015), Towards global partnerships: a principle-based approach to enhanced cooperation between the United Nations and all relevant partners*. https://undocs.org/A/RES/70/224

United Nations. (2017a). *Indigenous Peoples' Rights and the 2030 Agenda*. https://www.un.org/development/desa/indigenouspeoples/wp-content/uploads/sites/19/2016/10/Briefing-Paper-on-Indigenous-Peoples-Rights-and-the-2030-Agenda.pdf

United Nations. (2017b). *The United Nations Permanent Forum on Indigenous Issues*. https://www.un.org/development/desa/indigenouspeoples/wp-content/uploads/sites/19/2016/08/Indigenous-Peoples-and-the-2030-Agenda.pdf

United Nations. (2019). *Asia and the Pacific SDG Progress Report 2019*. https://www.unescap.org/sites/default/files/publications/ESCAP_Asia_and_the_Pacific_SDG_Progress_Report_2019.pdf

UNESCO Institute for Lifelong Learning. (2017). *Policy Brief 8*. https://unesdoc.unesco.org/ark:/48223/pf0000247569

World Bank. (2018). *Indigenous People*. https://www.worldbank.org/en/topic/indigenouspeoples

7

LEADERSHIP OF THE FUTURE, FOR THE FUTURE

An insight into a unique transformative learning programme for sustainability capability

Kate Harris

Introduction

As I finish this chapter, I am sitting on a plane next to a young Australian economist who has recently graduated from a leading Australian University. On his return from Central America, this talented economist – passionate about the planet and its people – has decided to go back to live in Mexico, "off grid," and to "check out of society." It saddens me to see such passionate individuals finding the challenge of necessary change too difficult, and deciding to remove themselves from the situation. The question that this and similar examples raise for me is: how can we best support our next generation to engage, and provide the leadership required for a sustainable future? How can we support them so they do not withdraw from this challenge?

In this chapter I will discuss one response to the social ecological challenges we face: the design and implementation of a programme in transformational leadership for sustainability that can provide a pedagogical framework for enabling an effective societal response. I will tell the story of the programme that I helped design and deliver. The Centre for Sustainability Leadership (CSL) Fellowship Program developed independently of any tertiary institution and graduated over 700 young sustainability leaders in 13 years through an accredited post-graduate programme. To me, the CSL experience was a powerful demonstration of human responsiveness, of felt need seeking out and discovering solutions, and a response to our felt needs and the felt needs of all who came to study with us.

Transformation learning for sustainability

Humanity is now facing grave ecological crises (Ehrlich & Ehrlich, 2013). These include rapid climate change, ocean acidification and the mass extinction of species.

As a consequence, and because of the difficulties our political systems are having in facing these crises, future generations, and their leaders, will be confronted by challenges like none that have been experienced before. The critical question is, how must we – and how will our youth – respond to these challenges, and what capabilities will enable them and us to design and implement local-to-global sustainable trajectories for the future wellbeing of our species and our planet? From my perspective, acquired in part through prior learning in social-ecological inquiry, we need effective educational frameworks to aid developing leaders to face, and then embrace, these challenges (Hathaway, 2017). We need leaders empowered to design the tools and processes required to implement effective scalable actions and solutions (Sipos et al., 2008).

To achieve this, we need leadership education that enables whole person learning through working towards the emergence of an authentic self. This sort of personal engagement, commitment and responsibility is the missing element in most considerations of leadership. And this requires a pedagogy that is innovative, creative and transformational (Netzer & Rowe, 2010).

> The term "transformative learning," is defined as learning that enables irreversible, profound, emancipatory change. It arises in our values, world views, beliefs, perspectives, understandings, and frameworks (or "meaning schemes") for imagining, thinking, designing, planning and acting. It contributes to our day-to-day living: our means of relating to the self, others, and the built and natural world. It is our "highest" level of learning because it absorbs us in our relationship to learning: content and meaning is always known and responded to contextually, through relationship.
>
> *(Taylor, 2008, pp. 7–10)*

Three streams of transformative learning are often distinguished:

- psycho-critical transformative learning
- social-emancipatory transformative learning and
- psychoanalytical transformative learning (Cranton & Taylor, 2012).

However, as Hill (in press) outlines,

> Despite many documents having hopeful titles, most primarily focus on critiques of the present situation, or on just a small part of the issue, with little or no content on what might be done to enable meaningful whole system change: from person to planet.[2] And most suggested changes tend to be fragmentary and adaptive.

So how do we as educators effectively prepare our future leaders for enabling transformative change that can lead to ecologically sustainable and meaningful futures?

Transformative Sustainability Learning (TSL) (Sipos et al., 2008, p. 74) is a more recent development. It integrates sustainability education and transformative learning to enable cognitive (head-cognitive), psychomotor (hands-psychomotor) and affective (heart-affective) domains of learning (building on the insights of Bloom et al., 1956; 1964) to provide experiences that can enable profound changes in knowledge, skills and attitudes for achieving ecological, social and economic justice.

The TSL approach is a helpful model to use when assessing transformative educational frameworks and it can be used to assess the learning framework of the CSL programme.

Perhaps the most helpful answer to effective transformative leadership lies in the guidance from Professor Hill, which outlines three critical elements to be included in transformative educational programmes. Hill was also a key influence in the CSL programme.

Hill argues that to be effective all learning experiences must:

- be unique to the individual's needs (taking into account content, time, place, modes of delivery and contextual relevance)
- nurture relationships (with the teachers – as mentors and guides – with other learners – collaborators in learning – and supporters, including family and community members)
- develop wisdom and ethics (being), as well as knowledge and skills (doing) (Hill, in press).

These elements were all central to the theory behind the practice that emerged in the CSL programme. In focusing on authenticity, and self-transformation in developing leaders, CSL incorporated insights from many areas.

The programme also integrated insights from restorative learning (Lange, 2012; 2015), theory U (Scharmer, 2007), conscious leadership in purpose-driven organisations (Brown, 2011; 2012), focusing (Gendlin, 1996), transformative arts-based learning (Lloyd, 2011, p. 155) and the connective work of Macy and Brown (2014). Each address self-awareness and emotional healing in a world understood as our larger living body (see also Fisher, 2002). In effect, CSL imagined transformative learning as part of the healing of our world.

CSL was, and is, not alone in teaching in this way. It is only one example of this sort of response. The significance of CSL lies in the fact that it was our programme, we designed it in response to felt needs, it represented and communicated the excitement of our learning in a world in transition and it was embraced by those who encountered it. As key participants in what we came to appreciate as a learning-based feedback system, we were enriched and continue to be enriched by this powerful learning experience. I will now describe its framework and curriculum as a pedagogical case study of transformative learning for sustainability.

Background

The Centre for Sustainability of Leadership is a not-for-profit organisation. It was founded in 2005 by a passionate 21-year-old woman, Larissa Brown,[3] as her response to deep concerns about the ineffectiveness of university systems in education for sustainable futures. She worked with Jason Clarke, founder of Minds at Work,[4] who became her mentor. He guided and facilitated her through the first few years of establishing the programme.

In preparation for this, Brown travelled the world and spoke with over a hundred leaders. She shared with them her concerns about the environment and climate change. She wanted to understand the strengths and skills they held in common. This informed the design of the core curriculum.

There was no funding in the beginning. It was a labour of love, run from a suburban bedroom. Teaching venues were donated and facilitators volunteered their time.

Participants were recruited through word of mouth, which is a somewhat simplistic way of describing the way in which passionate and purposeful individuals gravitated towards the commitment and enthusiasm the programme personified.

Within three years, the programme began receiving funding. It rode on the back of a political wave within Australia that spoke of carbon taxes and climate solutions. State and Federal Governments came on board with support. Unfortunately, this support was short lived: the politics changed. Subsequently, CSL sought and received core funding and placements through corporate sponsorship. However, the support of corporations was always problematic as many sponsored students began pursuing pathways beyond the corporate sector as a consequence of a paradigmatic shift that occurred through their learning.

My journey with CSL began in 2009, two years after completing my Masters of Social Ecology, and it was a baptism of fire. Designing the Sydney programme, only three weeks before it was to commence, was challenging, especially as there was minimal documentation of teaching done elsewhere. Our curriculum was designed with the support of a Melbourne facilitator, Stephen McGrail, under the guidance of Larissa Brown. The stated intent of the programme was "to get people who care, into positions of influence" because getting people in positions of influence to care had consistently failed.

We designed an eight-month fellowship programme that involved a weekly three-hour workshop and three week-long retreats. Objectives and key themes were designed to develop leaders of sustainability who would change the world for the better through their passion and their learned skills. The programme was organised around three equal principal phases, each with its own retreat. These have been described by Mah (2014, pp. 73–4) as:

Going within, through deep self and group reflection
Stepping up, through a focus on individual action and collaborative prototyping within projects and

Stepping out, through the implementation of iterative cycles of personal reflection on change in relation to collaborative real-world projects.

The core themes and outcomes used to support and further develop participants' capacity for change include *creativity, connectivity, capability* and *community*. Ways to grow these were integrated throughout the programme. *Creativity* was considered a key source for all effective solutions. We asked participants how they would "problem solve," and "problem avoid," using their new ways of knowing, being (including enjoying and enduring) and doing – we avoided managing, controlling and educating in a formal sense. *Connectivity* was initially introduced to connect one's creativity to a sense of purpose. However, it soon became a core focus throughout all three phases of the programme. Developing connection within the cohort was essential for the transformative nature of this programme. For developing *capability,* a list of ideal leadership skills was created. These comprised the competencies that were woven throughout the three core phases. To nurture *community,* it was made clear that participants were taking on a huge responsibility – essentially signing up to change the world. The cohort with whom each were experiencing this transformative process was identified as key to the success of the programme. Trust was paramount. If there was not the willingness to be courageous, open and supportive of one another, the trust of the process and of the group would not be realised.

The programme began with a three-day retreat within nature. This involved sharing, connecting, exploring and listening, as well as learning and honouring the stories of self and others. A key objective of this retreat was to establish trust – within participants themselves, with the facilitator, with the process and with each other. Without this foundation, the programme would not be able to achieve its potential. During the first retreat, participants would often cry with joy. They experienced an overwhelming sense of relief that they were not alone in their struggle to achieve a sustainable and equitable future. "I didn't know a community of people like myself existed." There was a feeling of finding home.

Establishing connections with one another was therefore key to the success of the programme – and it didn't just happen by chance. It was achieved through an intuitive sensing of who was right for the programme, when it was right for them to experience it and, importantly, who was not ready or fit for it. There were individuals who applied several times before being accepted, and many reported that this resulted in them doing the programme when they were best ready to get the most out of it.

The selection criteria considered:

- Readiness: to embrace new ways of knowing, learning and doing.
- Vulnerability: aware of and vulnerable to the self, the system and solutions to problems, and aware and vulnerable enough to work in a community of diverse experience and perspectives. (It was often noticed that there was commonly a lawyer, doctor, digital creator, campaigner, policy maker and researcher in each group.)

- Capability: having the potential to change the world, possessing the resilience and reverence to appreciate ecological systems, and being committed to creating change with care and consideration.

These qualities were further developed in the learning systems through the participants' attention to the interconnections and interrelationships within groups and between living systems (Johnstone, 2002). This was recognised as core to problem solving, which involved perceiving and appreciating the problem, prior to understanding and finding solutions. As a facilitator, my own connection to the unique potential of the individual was key to ensuring that learning outcomes were contextualised and meaningful. The sum of the whole, the group dynamic, was paramount to the success of the programme. My own self-awareness and connection to community was constantly tested. In a practical, rather than an egoic sense, my integrity was critical in the delivery of the platform.

In addition to the abovementioned theoretical educational and philosophical frameworks, Sara Parkin's (2010) work *The Positive Deviant: Sustainability Leadership in a Perverse World* was invaluable to the ongoing refinement of the programme, particularly her insights into what she calls the 4Rs: Reflection, Reverence, Relationships and Resilience. Retreats were held in physical environments chosen for their access to nature, providing quiet and space for deep introspection and creative vision. Reverence was nurtured through activities such as meditating and visioning, particularly during the first and last retreats. Relationships were more incidentally enabled; however, the relationships within groups were among the most valued elements of the programme and of the community experience. The importance of resilience was initially least appreciated, and it remains an area that requires greater attention. It is the key to long-term effectiveness. Interestingly, it was the skills of reflection, reverence and relationships that were key in supporting and enabling the development of resilience over time.

Processes: planned and emergent

Emergent learnings from the programme included the importance of attention to shared learning, language and experience. Although many shared learning experiences were intentionally provided – through coaching sessions, mentoring and project delivery – equally important were unplanned and spontaneous experiences.

I found it was amusing to witness the construction of shared languages within groups. These evolved as shorthand to describe core experiences, over the life of the programme. This shared language and experience became very important with the emergence of alumni groupings. And then I enjoyed watching conversations between new participants and alumni. The alumni seemed to appreciate exactly what participants were feeling in relation to the particular week of the programme that they were in.

Understanding the importance of providing and enabling shared experiences became more and more a priority over the life of the programme. Experiential learning and action-based learning experiences were also key to the programme. All sessions started with a grounding experience (such as using one or more of one's senses to connect to one's feelings and to the present environment), and an Acknowledgement of Country (paying respect to traditional owners and ongoing custodians of the land). These rituals are important anchors for both individuals and groups: they position values in relation to the living earth and to the learning. The valuing and integration of indigenous knowledge into the programme has increased over time. Significantly, alumni groups continue to carry this forward in their own gatherings.

Another important ritual has been the creation of a "vision board" or "collage of the future" to share the futures that participants want to create. These were created during the first retreat and they have been invaluable ever since. They use the crafting of messages and images to reach beyond just words: to transcend intellectual understanding. These vision boards are revisited at the end of each programme. Many alumni have told us that they still have their vision board and are astounded by the insights contained in it and its relevance to their experiences over time.

The outcomes of CSL

There are now over 700 CSL alumni. The overwhelmingly common feedback after completing the programme, year in year out, has been that "it has changed my life." Although our graduates are scattered across the globe, in many countries and in many roles, they all share a broadly similar vision of a future that can work. They know that they have the potential to contribute to the change required to achieve a sustainable future. They work in NGOs, government organisations, start-ups, community groups and many other structures and systems.

Examples of projects they have initiated include *Future Business Council, Wildwon, Our Say*, and *Kids in Nature,* each of which deserves a chapter of their own in a book such as this.[5]

As a sum of individuals, and as a collective, our graduates have had a significant impact. They have done this through their strategic and systemic understanding of what they do and how they do it. Many experienced learning that led to an immediate change in their career and sector. NGO staff moved to government, government to NGO, corporate to government and government to entrepreneurial start-up. Most, after changing their sector and/or role, have continued in these positions for the longer term. Some participants even left their employment to work solely for CSL. When I left my role as Sydney Director of CSL, the programme staff was almost entirely made up of CSL alumni. And, of course, the alumni have become a powerful self-sustaining community that generates learning and belonging aside from the formal institution of CSL. The skills acquired were supportive of one another's reintegration into a re-visioned world.

And yet many challenges still face CSL. Firstly, the organisation as we had known it for 13 years came to an end in 2018. Happily, it was passed into the hands of another highly regarded institution, one committed to maintaining its legacy: Monash Sustainable Development Institute (MSDI).[6] We welcome this, just as we recognise that, like many programmes that focus on paradigm shifts, the CSL programme needs to undergo its own transformation. Perhaps this change is a new beginning, not an ending.

Secondly, the financial viability of the organisation was always, and continues to be, challenging. Some of this was because of the difficulty of articulating and demonstrating "return on investment." This is always difficult with something that is transformative and allusive, something that can change the participants' lives for ever without any guarantee of what form that may take. Other challenges relate to the realisation of the difficulties and personal costs of being a sustainability leader.

Over time it became evident to me that greater preparation for resilience and self-care must be a core requirement of education for sustainability leadership. Of Parkin's 4Rs (resilience, relationships, reflection, and reverence), resilience has been the hardest to achieve. Jason Clarke (a co-creator of CSL) used to say to us, "you need to care less, to do more." Here he is suggesting that the challenge we faced was (and is) caring too much. I see truth in this. Fatigue and failure have also been common experiences amongst us. But I am grateful that at times of perceived failure and fatigue, the success and remedy has been the community itself, supporting us to keep going and to face future challenges.

Personally, I acknowledge that I am yet to master the art of self-care. There is so much to be done. Perhaps that is all we have to hold onto while creating change that we may not see the fruits of during our lifetimes. Perhaps it will be at the eleventh hour that our human potential and passion for one another and our planet comes to the fore, and transformational leaders will take the helm and boldly change what we do for the sake of the future.

Notes

1 http://www.csl.org.au
2 Four recent texts that Hill considers have made significant contributions to rethinking and redesigning our cultures are Hamilton (2017); Dale (2018); Washington and Towmey (2018); and McKibben (2019). These are cited and listed in the references in Chapter 3 of this volume by S. B. Hill.
3 Larissa Brown won the inaugural Australian Young Environmentalist of the Year award at the national Banksia Environmental Awards in 2008, and the Victorian Young Australian of the Year in 2010 (both recognising her roles in founding and leading the Centre for Sustainability Leadership). Since Jun 2018, she has been Director of Strategy and Policy for Senator Di Natale, past Leader of the Australian Greens: https://www.linkedin.com/in/brownlarissa/
4 http://www.mindsatwork.com.au/
5 *Future Business Council*: https://www.futurebusinesscouncil.com/; *Wildwon*: http://wildwon.com.au/; *Our Say*: https://home.oursay.org/; *Kids in Nature*: https://www.kidsinnaturenetwork.org.au/
6 Monash Sustainable Development Institute (MSDI): https://www.monash.edu/sustainable-development

References

Bloom, B., Englehart, M. Furst, E., Hill, W., & Krathwohl, D. (1956). *Taxonomy of educational objectives: The classification of educational goals. Handbook I: Cognitive domain.* Longmans, Green.

Bloom, B., Masia, B., & Krathwohl, D. (1964). *Taxonomy of educational objectives (Two volumes: The affective domain & The cognitive domain).* David McKay & Co.

Brown, B. C. (2011). *Conscious leadership for sustainability: How leaders with late-stage action logics design and engage in sustainability initiatives* [Doctoral dissertation, Fielding Graduate University, Santa Barbara, CA]. https://www.academia.edu/760595/CONSCIOUS_LEADERSHIP_FOR_SUSTAINABILITY_HOW_LEADERS_WITH_A_LATE-STAGE_ACTION_LOGIC_DESIGN_AND_ENGAGE_IN_SUSTAINABILITY_

Brown, B. C. (2012). Leading complex change with post-conventional consciousness. *Journal of Organizational Change Management*, 24(4), 560–577.

Cranton, P., & Taylor, E. (2012). Transformative learning theory: Seeking a more unified theory. In E. Taylor & P. Cranton, (Eds.), *The handbook of transformative learning* (pp. 3–20). Jossey-Bass.

Ehrlich, P. R., & Ehrlich, A. H. (2013). Can a collapse of global civilization be avoided? *Proceedings of the Royal Society B: Biological Sciences*, 280, 1–9.

Fisher, A. (2002). *Radical ecopsychology: Psychology in the service of life.* State University of New York Press.

Gendlin, E. (1996). *Focusing-oriented psychotherapy: A Manual of the experiential method.* The Guilford Press.

Hathaway, M. D. (2017). Activating hope in the midst of crisis: Emotions, transformative learning, and "the work that reconnects". *Journal of Transformative Education*, 15(4), 296–314.

Hill, S. B (in press). Transformative learning priorities. In D. Wright & S. B. Hill (Eds.), *Social ecology and education: Transforming worldviews and practices.* Routledge.

Johnstone, C. (2002). Reconnecting with our world. In A. Chesner & H. Hahn (Eds.), *Creative advances in group work* (pp. 186–216). Jessica Kingsley Publisher, John Wiley & Sons.

Lange, E. A. (2012). Transforming transformative learning through sustainability and the New Science. In E. Taylor, P. Cranton, & Associates (Eds.), *The handbook of transformative learning* (pp. 195–211). Jossey-Bass Publishers.

Lange, E. A. (2015). The ecology of transformative learning: Transdisciplinary provocations. *Journal of Transformative Learning*, 3(1), 28–34.

Lloyd, C. (2011). *Artful inquiry: An arts-based facilitation approach for individual and organisational learning and development* [Doctoral dissertation, Queensland University of Technology]. QUT ePrints.

Macy, J., & Brown, M. Y. (2014). *Coming back to life: The updated guide to the work that reconnects.* New Society.

Mah, J. (2014). *Supporting the journey of transformative learning and leadership for sustainability* [Master's thesis, Macquarie University]. Macquarie University Research Repository.

Netzer, D., & Rowe, M. R. (2010). Inquiry into creative and innovative processes: an experiential, whole-person approach to teaching creativity. *Journal of Transformative Education*, 8(2), 124–145.

Parkin, S. (2010). *The positive deviant: Sustainability leadership in a perverse world.* Earthscan.

Scharmer, C. O. (2007). *Theory U: Leading from the future as it emerges.* Society for Organizational Learning Press.

Sipos, Y., Battisti, B. T., & Grimm, K. A. (2008). Achieving transformative sustainability learning: Engaging head, hands and heart. *International Journal of Sustainability in Higher Education*, 9(1), 68–86.

Taylor, E. W, (2008). Transformative learning theory. *New Directions for Adult and continuing education*, 2008(119), 7–10.

8

THE GIFT OF PRESENCE IN GROUPS

An unfolding story of transformative learning

Dale Hunter and Stephen J. Thorpe

Introduction

Our universe is unfolding as a connected, networked and fully interdependent whole. This unfolding is holonomic, multi-layered and transcending time and space. Within this whole, in our world, the root systems of forests, the whole of human endeavour and the connectivity of the internet have much in common, as is recognised in the evolving discipline of social ecology.

Within this context, the threads we explore focus on a vital aspect of high-functioning groups and teams: that of being fully connected, "in the zone," accessing presence. Presencing can potentially become an integral part of transformative learning. Here we offer illustrative stories within which we are finding the words to describe *presence*, how and when *presence* occurs in groups, and how *presence* contributes to transformative learning.

What is this thing called presence?

> A group of people coming together in a state of *presence* generates a collective energy field of great intensity. It not only raises the degree of *presence* of each member of the group but also helps to free the collective human consciousness from its current state of mind dominance. This will make the state of *presence* increasingly more accessible to individuals.
>
> *(Tolle, 2004, p. 106)*

Light from the entirety of the night-time sky is present in every space – no matter how small. This is exactly the same phenomenon evident in a hologram. The three-dimensional image created by interacting laser beams can be cut in half indefinitely, and each piece, no matter how small, will still contain the entire image. This reveals what is perhaps the most mysterious aspect of parts and wholes: as physicist Henri Bortoft (1996) says, "Everything is in everything." When we eventually grasp the wholeness of nature, it can be shocking. In nature, as Bortoft (1996) puts it, "the part is a place for the presencing of the whole. This is the awareness that is stolen from us when we accept the machine worldview of wholes assembled from replaceable parts."

(Flowers et al., 2005, p. 5)

Deeper levels of learning create increasing awareness of the larger whole – both as it is and as it is evolving – that leads to actions that increasingly serve the emerging whole

(Flowers et al., 2005, p. 9)

Presence is conscious awareness. Collective *presence* and co-creation have been brought into the academic world through the work of Peter Senge and Otto Scharmer at the Massachusetts Institute of Technology (MIT). In *Theory U* Scharmer (2009) explains that:

> [Collective *presence* and co-creation] are empirically the rarest (generative forms of communication) and the most strategically important. That infrastructure, if in place, would allow whole eco systems to connect and cope better, faster and more innovatively with the key challenges at hand. The lack of that infrastructure represents a missing piece of societal hardwiring today.
>
> *(pp. 337–338)*

How and when does presence occur in groups?

Presence occurs naturally in focused and harmonious groups. *Presence* is sometimes called synergy, or group consciousness, and it is a form of collective intelligence. However, it is not collective intelligence associated only with the rational mind, though it may include this (see Woolley et al., 2015). Groups imbued with *presence* experience a vibrant field of aliveness, and each unique group will give this consciousness its own flavour, qualities and vibration. From sustained *presence*, unexpected and creative leaps and insights can emerge.

A conscious group will exhibit observable individual awareness, and individual and group *presence*. Consciousness grows where there is alignment, awareness and openness to more possibility. When a conscious group aligns sincerely on a purpose that is in keeping with the greater good of all, there is more energy available to the group. The group can become informed by a subtle energy that is greater than that usually available to individuals in the group. The form of this collective energy varies, and the strength of the energy seems to be related to the ability of the individuals in the group in generate *presence* and to the creation of a clear and aligned group purpose (and subsequently to effective action). The enhanced energy field of a group with an aligned group purpose can enable the individuals in it to raise their own individual vibration to meet that of the group. A group working consciously in this way attracts more consciousness to itself in the form of energy and information stored in the wider consciousness, called "Akasha" or the "Akashic Field" by Hungarian philosopher Ervin Lazlo (2004).

When this greater energy is consciously embraced and harmonised within a group, the group potential becomes more than the sum of the individuals within it. Irritations that may have previously been in the way fall into the background and are no longer as important or as relevant as they may have been. A noticeable shift occurs, from the constraints of individual ego into a fresh perspective and wider context of collective intelligence and then to a "higher self."

This effect has been described as synergy, flow or being "in the zone." The group becomes empowered or "powered up" and can work with the subtle energy of emergence.

In *The Art of Facilitation*, Dale Hunter (2007) explores how "dialogue in a space of deep listening opens up access to the whole field of consciousness, sometimes called the fourth dimension, and the unlimited knowledge available beyond the constraints of time and space" (p. 98).

In his book *The Living Classroom: Teaching and Collective Consciousness* Christopher M. Bache (2008) explores "learning fields" created between the teacher and class in his work as a university teacher. He uses the expression "collective abilities" in a similar way to our use of the term "collective intelligence."

> When people open themselves to each other and focus intensely on a common goal, their individual energies become synchronized in a way that can mediate contact with levels of intelligence and creativity that are beyond the reach of these same individuals acting alone. We must engage each other in an integrated manner for this more potent mode of knowing to emerge. The specific level of consciousness that is accessed is not as important here as the discovery of (... the enhanced capacity of the integrated group mind itself) ... Whatever our individual abilities, our collective abilities are greater.
>
> *(Bache, 2008, p. 68)*

Relationship with social ecology

Social ecology is the study of the relationships between people and their environments, including the interdependence of individuals, collectives and institutions (Hill, 2011). Social ecology recognises that nature is alive and connected. Humans are as much a part of nature (living things) as any other species: bees, trees, whales and all living cells. If we think and act as being separate from nature and from one another we lose our access to the *presence* that this connectedness brings. Awareness of the living reality of nature requires us to be in our bodies, as this essence cannot be accessed through intellect alone. All our senses and capacities are required to engage in direct experience.

Contemporary academia, with its diverse ways of learning, still embodies belief systems that are unsupportive of full human interconnection with the total living environment. This situation has its roots in the Renaissance, when a separation developed between religion and science as a practical way of allowing science to develop without the oppressive hierarchical control of the church. The illusion of pure objectivity, and the notion that scientifically provable forms of truth can be disconnected from the whole of living experience, is still dominant in parts of mainstream academia. This type of thinking does not allow for, or give credence to, diverse ways of knowing, including the direct experience of consciousness itself. Yet flashes of insight by our greatest mathematicians and scientists suggest moments of direct experience of unlimited consciousness.

As Stuart Hill wrote,

> Social ecology brings together so many poles that rarely meet: the arts and sciences; critical thinking, reflexivity, passion and intuition; rationality and spirituality; the stories of the ancients, systems theory and chaos theory; plus, an extensive list of disciplines. Our social ecology is a transdisciplinary meta-field that has been particularly informed by ecology, psychology and health studies, sociology and cultural studies, the creative arts, holistic sciences, appropriate technology, post-structuralism and critical theory, ecofeminism, eco-politics, ecological economics, peace and futures studies, applied philosophy and "spirituality" (in its broadest sense).
>
> *(Hill, 2011, p. 18)*

Presence fits into the broadest sense of spirituality. We prefer not to use that term though, as it might suggest to some that it belongs in some way to organised religion. It doesn't.

The transformative learning context

Although there are many educational theories that focus on the learner as an individual in a process of knowledge discovery (see Bengtsen, 2014; Bridges, 2017; and Moore, 2012), the group facilitation learning and facilitator education of Zenergy[1] can be located at a co-creation end or edge of transformative learning theory.

Transformative learning theory (Mezirow, 1990; 1991; 1994; 2000; Mezirow & Taylor, 2010) has, over the last few decades, remained true to its original construction. The theory has been described as constructivist, and "an orientation which holds that the way learners interpret and reinterpret their sense experience is, central to making meaning and hence learning" (Mezirow, 1994, p. 222). The theory involves two kinds of learning: *instrumental learning* and *communicative learning*. Whereas instrumental learning focuses on "learning through task-oriented problem solving and determination of cause and effect relationships" (Taylor, 1998, p. 5), communicative learning focuses on how others communicate their feelings, needs and desires with others. Awareness of transformative learning theory helps learners become more critical, autonomous and responsible thinkers (Mezirow, 2000). It has commonalities with other theories of adult learning, including andragogy (Knowles, 1984), experiential learning (Rogers, 1969), and with Cross's (1981) Characteristics of Adults as Learners model. In addition to these is an emerging third learning dimension: a *presence* and an awareness of collective intelligence providing access to seemingly limitless energy and knowledge.

This intangible quality of *presence* has been described by those who have recognised and valued this quality of being as essence, awareness, wairua,[2] the elixir, consciousness, healing, love, Gaia, and the interconnectedness of all things. Collective *presencing* is a transformative gateway through which holistic insights and creative solutions may emerge to address issues such as climate change, best use of scarce resources, wellbeing and equity. *Presencing* transcends traditional ways of thinking and acting and has the potential to enhance educational learning at all levels.

The Zenergy experience and transformative learning approach

Founded in 1993, Zenergy is a small globally active New Zealand-based facilitation company working towards the vision of "Whole people co-operating in a sustainable world." Zenergy runs training programmes with groups of people and works with organisations in designing and facilitating leading-edge group work practices.

Zenergy focusses on the role, skills and potential of the neutral or impartial group facilitator (International Association of Facilitators, 2004) who guides process and does not become involved in content. We envisaged the skills of group facilitation being made available globally to all levels of society, particularly to those who are disadvantaged and open to empowerment (Freire, 1972).

Together, we explored co-operative ways of working and began working both as group facilitators and facilitator trainers. Our training programmes for group facilitators are usually held over four or five days. They are highly interactive and holistic, and include movement, music, interactive processes, side-by-side coaching and deep sharing.

Over time the leaders experienced many occasions in which the groups facilitated made unexpected leaps and generated transformative group experiences. Reflection on these experiences led us to explore synergy, *presence* and collective intelligence. As a result, we developed a passion for understanding and helping provide access to group *presence* and synergy.

Essence story

In "The Essence of Facilitation" (Hunter et al., 1999), one of the five-day, stage 2 facilitator training group experiences is described:

> This was a wider network of facilitators five-day group training involving 17 people in March 1996. Different group members experienced a whole range of expanded awareness. The shift that occurred was from a group of separate individual identities into that of expanded individual awareness and group consciousness. Most of the people who took part in this experience have continued to experience a high level of connection and synchronicity since that time. Many of the participants have remained in connection with one another (till now) and are involved in life enhancing work.
>
> *(p. 111)*

An interesting aspect of this *presencing* experience was that on the third or fourth day we forgot to have morning tea and lunch, and only became aware of this at about 3:00 pm.

Over time, a body of knowledge developed, which was synthesised into five books and a modular Diploma of Facilitation, including an online module for a time. Participation in face-to-face and online learning experiences included more than 2500 people, in several hundred small group intensives and numerous interactive sessions at facilitator conferences in many countries around the world. Online facilitator training was part of this.

In addition, the availability of the Zenergy books through amazon.com help spread this group facilitation knowledge to the USA, Europe, Asia and Australasia. There have also been translations for China, Korea and Spanish-speaking countries.

The Zenergy approach focuses on enabling groups to improve their understanding of group dynamics, to establish an aligned group purpose and to develop a shared culture for working together and supporting whole personhood (Hunter et al., 1994; 1997; 1999; Hunter, 2007). At a group level, this approach regards the learner as an individual within a wider field of collective intelligence. Thus, it extends transformative learning theory beyond that of individual learning to group learning. Through a dynamically facilitated process, learners interpret and reinterpret their sense experiences within a collective group context.

Elite sports

Presencing is a natural phenomenon that can be found in many groups and teams, including elite sports. In a *New Zealand Herald* article by Chris Rattue following a rugby game, All Black member Beaudon Barret refers to what we call *presence* as "being in the zone."

> We're aware that teams typically drop off, especially at the end of the first half … It's hard to keep that intensity right up there. Naturally you fatigue, and so we challenge ourselves to just work that little bit harder to keep the ball alive. It worked tonight and it worked last week. It comes down to work rate and believing … When you're thought-free you're out there in the moment, just doing it. That's when I'm at my best as an athlete and it's

> probably the same for everyone else. You're just in the zone ... you're executing, you're all connected. You're all on the same level. The challenge as an athlete is how do you get there at the start of a game as often as you can.
> *(Rattue, 2018)*

Gaining access

Presencing is described in various ways, including awareness, consciousness, spaciousness, inner stillness, connection to the whole, to the divine, "all of it," being awake, in the flow, in the zone and synergy. *Presence* often occurs spontaneously as part of in-depth involvement with activities such as music, dance, movement, sport and being in nature. It can also be accessed through spiritual practices, including prayer, meditation and Buddhist-based "mindfulness."

But for those who do not know *presence*, and are not involved in the above activities, where can they experience it and learn how to access it? And for the many who struggle with being in the world, and perhaps suffer from depression, this access could be crucial to finding well-being.

An important way to access *presence* and grow this capacity is to notice when it is present in a group. For many, it is a lot easier to notice *presence* with others than on one's own. This is because a group can act as an amplifier of *presence*. Being in nature and experiencing "awe" can have a similar effect. So being in a group, and in nature, is even better.

The need for a safe container

The provision of a safe "container" is important for groups working effectively and accessing *presence*. In our practice as facilitators we have found that the setting of an aligned group purpose and agreeing on a group culture (how we want to be together), sometimes known as "ground rules," is very important in creating safety. Purpose and culture provide a safe container for *presence*, consciousness and collective intelligence to emerge.

As a process it is important to allow participants to first become aware of and connected energetically with their physical body, to connect their energy with the earth and then to tune in with others around them and then beyond the room to the wider environment. This energetic connection can be also extended out to a planetary and cosmic level that is free of time and space. Then notice what emerges.

Here are some pointers to accessing *presence*. Adapt them to your own needs.

- Notice your breath moving.
- Feel into your whole body (perhaps starting with the hands).
- Feel your feet connect with the floor or ground.
- Allow energy and awareness to flow into the earth.
- Become centred in your body.
- Become aware of the whole room and beyond, including the natural environment.

- Notice everything that comes to you – thoughts, feelings, sensations – and let them fall away as you breathe out.
- Notice and be with the others in the room – listen deeply.
- Together "hold the space" for meeting the purpose of the group.
- Allow for emergence – speech, movement, images, colours, surprises or whatever shows up.
- If the energy feels a little heavy, it is helpful for the facilitator to maintain a degree of lightness and spaciousness – "like a feather on water."

Kāren's story

Here is an example from musician and facilitator Kāren Hunter:

> I was invited as a facilitator onto the team of a Transformational Festival for its final New Year event after a 24-year run as a central pillar of a nation-wide community. The festival catered for between 1500–2000 at each New Year's event, and I had noticed that in the aftermath of each event there was a high ratio of dissatisfaction between crew members. It had a tendency to become a "toxic" stressed working environment, which was the antithesis of what I expected. I didn't understand how the group could give so much joy to the community and yet come away themselves feeling completely drained.
>
> I told the group I was only interested in taking the role if a few members of the team could attend a Zenergy Stage One training programme and get some training to support me. Two key women took it on.
>
> Once they had completed the training, and there were now three of us who understood some basic processes and tools; we introduced the concept to the rest of the core crew, and we worked with them to create a Purpose and Culture for the festival to guide the final event. The group chose as the Purpose: *To nourish energetic sustainability.* This was in essence a group in crisis, and the purpose reflected the need to survive the final event! The Culture items were:
>
> - Trust
> - Compassion
> - Honouring
> - Team spirit
> - Communication
> - Self responsibility
> - CELEBRATE!!!
>
> Every festival involves high energy, problem solving, urgent communication and "number eight wire" mentality (a phrase used to describe a stereotypical mentality of New Zealanders referring to their creativity and the ingenuity). By being grounded and present to our purpose and culture, the team emerged from the event tired, but able to hold themselves and each other with respect

and tenderness. Festival participants had commented on the abundant "nourishing energy" within the crew, and although there were inevitable bumps and bruises the overall mood was upbeat and positive.

Sustainable co-operative processes in organisations

As part of my (Dale) PhD research in the Social Ecology Research Group (SERG) between 2000–2003, supervised by Emeritus Professor Stuart Hill, I facilitated a co-operative inquiry into *sustainable co-operative processes in organisations* (Hunter, 2003) which we had termed "co-operacy" (Hunter et al., 1997). Eleven people (most of whom were facilitators) took part over a number of weeks in this project, which involved individual research and four interspersed full-day meetings of the research group. On the last full-day session, the research group realised that co-operacy was not only about bringing about structural and process change in organisations (doing things differently). We had a transformative experience as a group in which we realised that – for co-operative processes to be sustainable – each of us needed to transform ourselves into fully embodied "whole persons" imbued with love and compassion. This insight emerged during a sustained period of several hours in which we were *presencing* together (Hunter, 2003).

The occasional spontaneous appearance and experience of *presence* in groups created a desire to learn how to generate this "power" consciously in groups as a facilitator. This journey of discovery has led to much learning about the nature of *presence* and its potential to transform or "supercharge" intentional groups into potent forces for good.

As a group participant, I found that the strong experience of *presence* generated within an aligned group, dimmed or disappeared after a group ended and I returned to "normal." To have more access, I learnt to recreate *presence* by myself, and did this through various techniques including writing a journal, going to the gym, practicing yoga, meditation, and using breathing techniques. Eckhart Tolle's videos helped me through pointers such as "saying yes to every moment," awareness between thoughts and spaciousness.

Presence works online too

Online facilitator training has also been a part of Zenergy's exploration, as we recognised the potential for generating collective intelligence in groups using emerging online group tools and technologies. Drawing on their face-to-face facilitation training programmes, Zenergy launched its first Online Facilitation Skills programme in 2006. The programme explored facilitation in online groups using text, audio, video conferencing and 3D software tools. As an inquiry, we wondered whether it was even possible to facilitate and create an environment for generating collective intelligence in online group settings, given the lack of "embodiment" (Hunter, 2003). Without loops of communication feedback, body language cues and verbal tone, there was a curiosity to uncover whether a facilitator could effectively create a safe container, build relationships and facilitate transformative group experiences as effectively online (Thorpe, 2011). Despite the various challenges of geography, different time zones and technology-mediated communication, we found that it was possible to take an online group to similar depths of connection and engagement as had been experienced in face-to-face facilitated groups, and that transformative group experiences were possible.

Another clear outcome was the development of a set of 26 online facilitator competencies, grouped into seven categories (Thorpe, 2016). Along with capabilities in group process and competence with online software tools, we recognised the need to "communicate with *presence* online." This important competence was articulated as the ability to facilitate online groups at a deeply creative and generative level. It involved the facilitator having the ability to *presence* self and others separated by time and distance, while being sensitive to cultural differences. "The facilitator was expected to hold and support a group through deep listening and careful communication, assisting the group through their interventions to harness their collective intelligence and achieve their best performance" (Thorpe, 2016, p. 84).

Rick's story

Facilitators trained in Zenergy methods took their knowledge and skills back to their own workplaces. One Zenergy-trained facilitator and master electrician, Rick Sommerford, worked for a number of years (2014–2017) on the development of a large US$37 billion Liquefied Natural Gas (LNG) construction project near Darwin (Australia). Rick was leading a team of 55 electricians and trade assistants, initially employed on night shift to lay the cabling for the project. Safety and productivity were important issues.

Each shift began with a briefing meeting called "prestart." The purpose was to inform the team of specific tasks for the night, raise safety issues and build relationships within the team. Rick and his colleague, Scotty Baker, led these meetings. Rick regarded the opportunity to build relationships within the team as a way of increasing safety and productivity. He introduced a group process into this meeting in which one of the team (a different person at each shift) was invited to share something about themselves unknown to his or her co-workers. Interesting and personal information was shared: at times amusing, at times poignant and deeply felt. As the project continued over eight months, this sharing became an important part of the meeting and it led to a deepening of relationships and a strengthened sense of team work. They also trialled group awareness games to focus the team on each other and their interconnectedness, and on how different mindsets can influence the group dynamic.

Rick says,

> at times the sense of *presence* was palpable. Productivity increased beyond what is normally associated with nightshift. For example, a 0.8 efficiency rate for nightshift is accepted as a good result for equitable dayshift activities. On many

> occasions the night shift team achieved efficiency rates equal to 1.2 efficiency of that achieved on dayshift. This is extremely rare in the industry.

There were comments around being "in the zone tonight" or "wow, how did that happen" and "a knowing in the group, that something special had happened, that something I had come to know as *presence* through Zenergy had been there that night." Team social events occurred spontaneously. At the end of a Saturday night shift with a night off on Sunday, the team members would head off to party a little and enjoy the fruits of their teamwork.

This way of experiencing *presence* in a team environment was the most satisfying experience at work that many of the group had experienced, including Rick and Scotty.

Zenergy leaders' meeting

The work is continuing. At the latest five-day retreat, attended by seven leaders, the following experience was noted:

> The purpose of the retreat was "To refresh and refocus the work of Zenergy in the world." On the fourth day, the energy appeared to rise up in the centre of the group like a "spout." The energy felt so strong that we found ourselves standing up and moving further apart. We shared our impressions and thoughts as we "held" the energy. When the energy subsided after an hour or so, we drew a picture jointly of a whale emerging from spouting water. This was the closest we could come to representing our experience. The experience had the effect of enlivening and energising us to move into the next phase of development as a self-motivating collective.

Conclusion

The ability to generate *presence* is part of developing effective groups. However, presence cannot be imposed or forced. It is a natural phenomenon that surfaces when a group is aligned, has a clear and life-enhancing purpose, a supportive group culture, and the commitment to do the work. *Presence* and conscious awareness in groups is a "super power" that is gradually being recognised, accessed and named (although not always with the same name). *Presence* works beyond time and space, and can lead to unexpected and, at times, remarkable outcomes and actions for the common good. *Presencing* can become a highly valued part of cooperative transformative learning.

Notes

1 https://zenergyglobal.com/
2 Wairua is a Māori word that may be translated as "spirit"; a more extensive definition is provided at https://maoridictionary.co.nz/search?&keywords=wairua

References

Bache, C. M. (2008). *The living classroom: Teaching and collective consciousness*. State University of New York Press.

Bengtsen, S. S. E. (2014). Into the heart of things: Defrosting educational theory. In: P. Gibbs, & R. Barnett (Eds.), *Thinking about higher education* (pp. 175–192). Springer.

Bridges, D. (2017). *Philosophy in educational research: Epistemology, ethics, politics and quality*. Springer.

Cross, K. P. (1981). *Adults as learners*. Jossey-Bass.

Flowers, B. S., Jaworski, J., Scharmer, O., & Senge, P. M. (2005). *Presence: Exploring profound change in people, organizations, and society*. Doubleday.

Freire, P. (1972). *Pedagogy of the oppressed*. Penguin.

Hill, S. B. (2011). Social ecology: An Australian perspective. In D. C. E. Wright, C. E. Camden-Pratt, & S. B. Hill (Eds.), *Social ecology: Applying ecological understanding to our lives and our planet*. Hawthorn.

Hunter, D. (2003). *The facilitation of sustainable co-operative processes* [Unpublished doctoral dissertation]. The University of Western Sydney.

Hunter, D. (2007). *The art of facilitation. The essentials for leading great meetings and creating group synergy*. Random House.

Hunter, D., Bailey, A., & Taylor, B. (1994). *The art of facilitation: The essentials for leading great meetings and creating group synergy*. Tandem Press.

Hunter, D., Bailey, A., & Taylor, B. (1997). *Co-operacy: A new way of being at work*. Tandem Press.

Hunter, D., Bailey, A., & Taylor, B. (1999). *The essence of facilitation*. Tandem Press.

International Association of Facilitators. (2004). *Statement of values and code of ethics for facilitators*. https://www.iaf-world.org/site/professional/iaf-code-of-ethics

Knowles, M. (1984). *Andragogy in action*. Jossey-Bass.

Lazlo, E. (2004). *Science and the Akashic Field: An integral theory of everything*. Inner Traditions.

Mezirow, J. (1990). *Fostering critical reflection in adulthood: A guide to transformative and emancipatory learning*. Jossey-Bass.

Mezirow, J. (1991). *Transformative dimensions of adult learning*. Jossey-Bass.

Mezirow, J. (1994). Understanding transformation theory. *Adult Education Quarterly*, 44(4), 222–232.

Mezirow, J. (2000). *Learning as transformation: Critical perspectives on a theory in progress*. Jossey-Bass.

Mezirow, J., & Taylor, E. W. (2010). *Transformative learning in practice: Insights from community, workplace, and higher education*. Jossey-Bass.

Moore, T. W. (2012). *Educational theory: An introduction*. Routledge.

Rogers, C. R. (1969). *Freedom to learn*. Merrill.

Rattue, C. (2018, August 26). Why the All Blacks will win the World Cup. *New Zealand Herald*. https://www.nzherald.co.nz/sport/news/article.cfm?c_id=4&objectid=12113802

Scharmer, C. O. (2009). *Theory U: Leading from the future as it emerges*. Berrett-Koehler.

Taylor, E. W. (1998). The theory and practice of transformative learning: A critical review. *Information Series, No*. 374. Center on Education and Training for Employment.

Thorpe, S. J. (2011). Using storytelling in the facilitation of online groups. *Group Facilitation: A Research and Applications Journal*, 11, 24–39.

Thorpe, S. J. (2016). Online facilitator competencies for group facilitators. *Group Facilitation: A Research and Applications Journal*, 13, 79–90.

Tolle, E. (2004). *The power of now: A guide to spiritual enlightenment*. New World Library.

Woolley, A., Arggarwal, I., & Malone, T. W. (2015). Collective Intelligence and group performance. *Current Directions in Psychological Science*, 6(24), 420–424.

9

ART, IMAGINATION AND THE ENVIRONMENTAL MOVEMENT

Rachael Jacobs and Christine Milne

Creativity and change

Australia's most influential environmental activist Bob Brown likened the flooding of the Franklin River to "putting a scratch across the Mona Lisa" (cited in Press, 2018). This accords with Hennessey and Amabile's (2010) argument that "it is only with creativity that we can hope to address the myriad problems facing our schools and medical facilities, our cities and towns, our economy, our nation, and the world" (p. 570). Writing this chapter in 2019, as the ash smog from a recent outbreak of forest fires swirls overhead, the issue of the climate emergency permeates every consideration of our economy, our nation, and the world. At the time of writing, 11,000 scientists have warned that the "climate crisis has arrived and is accelerating faster than most scientists expected," which will bring people to face "untold suffering" unless major transformations are made to global society (Carrington, 2019). The scientists go on to explain that it is not too late to take action, and the article subsequently reports that the world is currently witnessing a surge in activism around the climate crisis. Aside from public activism, communities are increasingly making changes to personal behaviours, such as using less plastic, and flying and driving less. There are calls for companies to publish their environmental credentials and be held accountable for any environmental damage they create. Bodies of governance all over the world – from local councils to the EU – are declaring a climate emergency (Rankin, 2019) in the hope of pressuring governments to transition energy sources to renewables, to strengthen protection of vulnerable areas and to mitigate the effects of the climate crisis.

To further mobilise communities into action, aesthetic modes of campaigning and activism must be used to communicate the plight of the planet. Creative thinking and imagination are also necessary to find solutions to the climate emergency that are intersectionally just, and that value all people and all beings as critical contributors to the ecosystem. As prominent educational and arts philosopher Maxine Greene (2009) said: "We need imagination. We need visions of the possible and of the unexpected" (p. 398).

Griffin (1996) argues that social movements come from a change in perspective of the population as "a change in public perception will change the public" (p. 65), and Greene (2007) adds that art can trigger a "transformative moment of moral and political awakening" (p. 2). Gablik (2004) also argues that art has a role in changing humans' worldview, as it challenges people to consider their role in the world outside of their immediate interests, wants and desires. Art can help us to participate in what Thomas Berry (1999) has deemed the "great work" of our time: "moving from a devastating presence on the planet to a more benign presence" (p. 62). Moon et al. (2013) also discuss Greene's concept of *social imagination*: in contrast to the popular perception of imagination as an individual process, social imagination "attempts to bring a personal process into a public space" (p. 231). They go on to highlight the importance of imagination and the arts in awakening a consciousness of social justice, stating that "social imagination is the starting place for creating a different society" (p. 232).

Curtis (2017) describes "three pathways through which the arts promote pro-environmental behavior" (p. 4). The arts can be used to communicate ideas to audiences, the arts connect audiences to nature, and the arts can embed ecologically sustainable development through artworks. In this chapter we focus on the first two of Curtis' pathways. Although we recognise that ecological art is a growing and critical movement of arts activism, we do not engage in a detailed discussion of its implications for the environmental movement; rather, we direct the interested reader to the writings of Curtis (2017), Marks (2017), Carruthers (2006) and Wallen (2012). We also do not engage in a critique of mainstream art practices, which are often criticised for their competitiveness, individualism and profit-making (Costantino, 2011; Gablik, 2004; Wallen, 2012), these being patterns of behaviour that are contra to the ideals of climate justice and environmental protection. Rather we reflect on flashpoints in Milne's (2017) renowned history of environmental activism that have been shaped and influenced by the arts. These flashpoints will be discussed in relation to their influence on public opinion, interrogating the extent to which art and artists can make a difference to the global environmental crisis.

Miles (2014) argues in *Eco-Aesthetics: Art, Literature and Architecture in a Period of Climate Change* that beauty, which is "radically other to routine" (p. 36) can re-inflect today's culture and bring us to higher understandings of ways to make positive changes in community. The environmental movement is filled with examples of the ways that beauty has become a touchstone for social activism. In 1972, in Tasmania, "the heart of the south-west wilderness was lost when Lake Pedder was flooded" (Milne in Burgess, 2019). The engineered flooding resulted in the loss of a magnificent beach, a unique natural heritage site, and a place of significance for Tasmanians and Australians. The pink quartzite beach captured the imagination of Australians and people around the world as a result of the work of photographers and artists. The 1970s were days of black and white television. Some of the "the Sunday Painters," watercolour painters – including Max Angus and Patricia Giles – came to Lake Pedder to create striking full-colour images of the lake for the world. Olegas Truchanas was a wilderness photographer who captured the beauty of the lake through slideshows. These slides were shown in town halls and city halls all over Tasmania and mainland Australia, advertising the beauty of Tasmania and contributing to its tourism culture and economy. Sadly, Truchanas' photographic collection was destroyed in the Tasmanian fires of 1967, and he returned to try to re-photograph those areas (Angus, 1975). These artists, and many more, were a critical part of the campaign to save Lake Pedder. The current campaign to restore Lake Pedder (of which Christine Milne, a co-author of this chapter, is co-convenor) is informed by the strong presence of artists in past campaigns, and maintains a strong artistic tradition.

Before the advent of colour television, people were keenly interested in colour photography. Peter Dombrovskis (2017) followed the tradition of capturing the beauty of Tasmania for the purposes of environmental protection. His photograph of the of Rock Island Bend on the Franklin River was one critical photograph that became the symbol of the campaign to save the Franklin from damming in the early 1980s. Many Australians instantly recognise the image of the misty waterway dramatically swirling around ancient rock formations framed by lush foliage. Dombrovskis made the unique and untouched beauty of the Franklin River accessible to people who had never been to, nor were ever likely to visit, Tasmania. He mass-produced these images as calendars, which gained popularity in the homes of environmentalists and non-environmentalists alike. The campaign to save the Franklin was able to gain support from around Tasmania, and all over mainland Australia, from everyday people who felt that its beauty could not be sacrificed for short-term gain, as Lake Pedder had been. These two campaigns also began a tradition of wilderness photography in Australia, which is now a thriving community of artists (Dombrovskis, 2017).

The campaign to save the Franklin was also strongly supported by musicians. Folk artist Shane Howard of the Australian band Goanna wrote songs for the Franklin at the time of the campaign. While touring in 2013 with Carole King, he played the campaign's anthem "Let the Franklin Flow" to audiences who instantly recognised the song. Appreciative audiences sang along passionately to the chorus, declaring loudly that the song's central theme "there has to be something worth fighting for" is still relevant today. Similarly, campaigns to save Tasmanian wilderness have been supported by a rich arts literature. Kevin Kiernan (2017) wrote a compelling piece about the flooding and loss of Lake Pedder, and Richard Flanagan (2007) wrote an impassioned piece on the horrors of logging in Tasmania which was published in the British *Guardian*, then in Australia in *The Monthly*. The convergence of visual art, music and literature suggest that the rich intertextuality of the Franklin campaign was able to capture the imaginations of a wide range of Australians who were able to find an affinity and passion for places where most of them had never been.

In light of the convergence of artists in environmental activism in the 1970s and 80s, and as a result of watching the tremendous capacity for change that artistic engagement presented, this chapter's co-author, Christine Milne, re-developed The Tasmanian Greens' Arts policy in 1989, declaring that "The arts is the key to transforming Tasmania" (Tasmanian Greens, 1989 in Milne, 2017). This policy was an acknowledgment that since the 70s the artist has been key in re-thinking Tasmania as a place of natural beauty and protection, rather than maintaining a focus on resource extraction and wealth for extraction companies.

There are other examples outside of Tasmania demonstrating the success of campaigns that connect through the aesthetic senses, often using a kind of environmental iconography to drive a message of protection. In the campaign to protect the Daintree Rainforest, the largest continuous area of tropical rainforest in Australia, the Australian Conservation Foundation used the white lemuroid ringtail possum (*Hemibelideus lemuroides*) as the symbol of the campaign. The animal's image became iconic as the symbol of the wider issue of the protection of forests. Renowned French jewellery maker Lea Stein collaborated with Christine Milne and Sandra Harding, the Vice Chancellor of James Cook University, to produce the white lemuroid ringtail possum brooch. Stein's work is usually whimsical and imaginative and this was her first work highlighting a conservation issue. Stein, who is now a recluse in her 80s and only makes two designs per year, had to be approached in person and provided with the research around the endangered nature of the species, along with accompanying photographs. In Milne's (2017) book *An Activist's Life*, she details the artistic nature of the brooch, which is a collaboration between four women of politics, academia, artistry and small business. The species was strategically chosen. The white lemuroid ringtail possum was once predicted to be the first extinction driven solely by global warming. This position has been sadly taken by the Bramble Cay melomys (*Melomys rubicola*) – last seen in 2009 – but the white lemuroid ringtail possum is still endangered.

Aesthetic campaigning and aesthetic education in the contemporary climate crisis

There is an undeniable shift of perspective when one's aesthetic senses are engaged. This can encourage one to question and critique the state of things, and even spur one to action. In *Art as Experience*, Dewey (1934) advocated that conscious involvement with aesthetic experiences provides a marked difference to everyday experiences. However, environmental campaigns that focus primarily on the science (e.g. of global warming) have dominated, particularly over the last decade. It is unsurprising that the environmental movement has invested so heavily in turning public opinion just through facts, figures and science. From an early age, we implicitly learn that rational and functional modes of thinking are more commonly privileged in Western education systems, and aesthetic values are often absent in discussions about learning and development. Martin-Smith (2005) argues that this tension originated from Descartes' philosophy that dissociated the mind from body and considered aesthetic feelings to be associated with irrational senses of the body rather than reasoned thought. There may also be a level of discomfort around what the arts reveal about ourselves and society when exposed. Baldwin (1962) states:

> The state of birth, suffering, love, and death are extreme states—extreme, universal, and inescapable. We all know this, but we would rather not know it. The artist is present to correct the delusions to which we fall prey in our attempts to avoid this knowledge. It is for this reason that all societies have battled with the incorrigible disturber of the peace—the artist.

The discomfort is often exacerbated by Western education systems which routinely diminish the importance of aesthetics in dialogues in schooling and curriculum. Aesthetic education carries common perceptions of highly ethereal qualities that are abstract in nature or often associated with high culture (Ross et al., 1993). The importance of aesthetic education is further diminished when aesthetic is perceived as involving private and "feelingful" responses that are only engaged with on an individual level. Aesthetics are commonly seen as intangible, similar to the concept of "creativity," which we are used to thinking about in romantic terms of a "talented" individual working alone to produce highly original work.

Writers in arts education (Attwood, 2015; Jacobs, 2009; Dunn, 2005) have long explored the importance of aesthetic literacy in the school curriculum. Although an unfamiliar concept to some, the term "aesthetic literacy" is not new. In 1978, Greene used the term to describe a kind of sensibility that can be arrived at, or a form of "conceptual awareness" that enables "diverse persons to break through the cotton wool of daily life and to live more consciously" (Greene, 1978, p. 185). Greene (1999) later reaffirmed this vision, saying

> sometimes I think that what we want to make possible is the living of lyrical moments, moments at which human beings (freed to feel, to know, and to imagine) suddenly understand their own lives in relation to all that surrounds.
> *(p. 7)*

It is a vision of learning that cuts across disciplines, presenting a core value that deserves more attention in contemporary curricula. Allowing students to become more aesthetically aware at school provides them with a way of seeing that transcends the instrumental and disciplinary approaches where we are able to "learn from aesthetically rendered lives what words, paradoxically, can never say" (Eisner, 1985, p. 35).

This learning can take place outside of formal schooling. In some ways, artists, arts educators and arts activists have a vital role in this regard, as they are able to work outside of the restrictive boundaries of curriculum and educational policy. In fact, aesthetic concepts and the poetic language used to describe them have become alienating to some curriculum and policy writers, who are more accustomed to programming learning that facilitates rational and functional modes of thinking. It is difficult to explain to teachers and policy writers that they should be concerned with "the living of lyrical moments." Additionally, aesthetic education in a Western context is often considered as academically "soft" or less rigorous than other ways of knowing. All of these factors have combined to create a schooling system with a degree of comfort in facts and figures, evidence-based knowledge, deductive reasoning, critical analysis and rational thought. The work of artists, arts educators and arts activists is able to have impact beyond the classroom, using inter- and intra-generational dialogues regarding the care of the Earth.

Both within and outside of the boundaries of schooling, aesthetically charged experiences are important in environmental education, which is an area that has come to be defined by scientific proof and positivist knowledge. The science of the climate crisis is important, and scientists who conduct research into the issue feel a moral imperative to warn us of the threat of inaction. As scientists tell us about global warming, and the mass extinctions to come, it is easy to feel despair and to avoid having to deal with the issues by distracting and distancing ourselves. Author and futurist Steffan (2014) further warns us that politicians seem to believe that "expressions of concern and extremely modest, almost symbolic, small steps and half measures are the appropriate course of action." However, he goes on to declare that optimism and, indeed, hope are powerful political acts. Years previously, Bookchin (1991, in Best, 1998) argued that:

> … the ecological movement will never gain any real influence or have any significant impact on society if it advances a message of despair rather than hope, of regressive and impossible return to primordial human cultures and sensibilities rather than a commitment to human progress and to a uniquely human empathy for life as a whole.
>
> *(p. 334)*

It is worth mentioning that the scientists warning of the dangers facing humankind from the climate emergency have emphasised that it is not too late to change course to avoid catastrophic consequences (Carrington, 2019). Environmental scientists have also had the indescribably difficult job of countering climate denialism, using their collective voices to counteract the loud microphones that climate sceptics have been handed. However, the collective voices of the science community can only go so far in addressing this situation. To spark a critical mass of people into driving change, the aesthetic sense must also be engaged. Everyday people's humanity can be activated through empathy, and they need to imagine new, wild possibilities for a world that is cleaner, more sustainable and ecologically just. The engagement of art in this quest is critical. Greene (2007) reminds us that art can trigger a "transformative moment of moral and political awakening" (p. 2). It is possible to capture the world's attention through an encounter with a text that need only take a moment to process. If we are imaginative when we encounter such artworks, "our way of seeing the world widens; we see in a different light" (p. 2). Griffin (1996) goes so far as to suggest that "social movements are driven by imagination" (p. 65).

Motivating action through knowing and feeling

In the discussion between the authors used to theorise this chapter, Milne illustrated the difference between a scientific argument for climate action and an aesthetic artefact using the analogy of two television shows. An episode of *Catalyst* (an Australian Broadcasting Corporation series featuring science stories from around the world) activates the viewer's rational sense, giving them facts and figures, background information, theories, findings and implications of the research. In comparison, a David Attenborough documentary captures the beauty, wonder and awe of the world that ignites the human spirit. The information communicated on *Catalyst* is important, but it does not adequately touch the heart. If the environmental movement becomes obsessed by a *Catalyst*-style fact-rich method of communication that aims to reason people into climate action, they deny a major motivational driver of what causes people to act. That major motivational driver is a sense of lyrical feeling, connection and awe that can be created and nurtured through the arts.

As a dancer, Jacobs is currently working with the Extinction Rebellion movement to choreograph activist gatherings, making them visually spectacular, joyful, engaging and inviting for a wider audience. While the medium of dance is exciting for the participant, this style of participatory engagement in arts activism has the dual effect of releasing the power of the imagination in its audience (Greene, 2007). Passers-by pull out their phones to record the spectacle. While the movement aims to disrupt, a sense of sympathy is activated for the protesters as they demonstrate "the capacity to break with the ordinary, the given, the taken-for-granted and open doors to possibility" (Greene, 2007, p. 1). The movement becomes hopeful, optimistic and accessible, allowing others to imagine the better world that the activists wish to create. In that moment, the sense of hopelessness is lost as art becomes a mirror that is able to capture what is so difficult to put into words. Furthermore, it becomes evident that "imagination lights the slow fuse of possibility" (p. 2). In this way, imagining is a revolutionary act. There are revolutionary possibilities in strengthening the muscle of imagination and reclaiming the right to imagine better futures.

Greene metaphorically states that "without imagination you live in a small room with the windows closed, imagination opens windows and shows horizons, and landscapes" (in Jefferson & Anderson, 2017, p. 81). It is the eco-artists' wish that those windows open onto landscapes restored by climate action, with clear horizons accessible to all. The transformational powers of the arts must be harnessed if we are to organise ongoing climate action beyond what is in our obvious and immediate world. Climate action must take place beyond mitigation, after the bushfires, on cold days as well as hot, and in regions near and far from those we are trying to protect. Through the arts and imagination, we can travel beyond our own words and reach beyond the common boundaries to create the mass movement of humanity that is so desperately required to create meaningful change.

References

Angus, M. (1975). *The world of Olegas Truchanas*. Olegas Truchanas Publication Committee & Harper Collins.

Attwood, A. I. (2015). *Aesthetic literacy through the avant-grade: Establishing an aesthetically responsive curriculum* [Unpublished doctoral dissertation]. Washington State University.

Baldwin, J. (1962). The creative process. *The open space of democracy*. https://openspaceofdemocracy.files.wordpress.com/2017/01/baldwin-creative-process.pdf

Berry, T. (1999). *The great work: Our way into the future*. Harmony/Bell Tower.

Best, S. (1998). Murray Bookchin's theory of social ecology. *Organization and Environment*, 11(3), 334.

Burgess, G. (2019, October 4). Draining Lake Pedder 50 years on gains environmental momentum. *ABC Radio Hobart*. https://www.abc.net.au/news/2019-10-05/campaign-to-drain-lake-pedder-ramps-up/11560808

Carrington, D. (2019, November 5). Climate crisis: 11,000 scientists warn of 'untold suffering'. *The Guardian*. https://www.theguardian.com/environment/2019/nov/05/climate-crisis-11000-scientists-warn-of-untold-suffering

Carruthers, B. (2006). Art, sweet art. *Alternatives Journal*, 32(4–5), 24–27.

Costantino, T. (2011). Contemporary aesthetic theory and models of creativity in visual arts education in the United States. In J. Sefton-Green, P. Thomson & K. Jones (Eds.), *The Routledge international handbook of creative learning* (pp. 159–167). Routledge.

Curtis, D. (2017). Building sustainability with the arts: Introduction. In D. Curtis, (Ed.), *Building sustainability with the arts: Proceedings of the 2nd national ecoarts Australis conference* (pp. 3–18). Cambridge Scholars Publishing.

Dewey, J. (1934). *Art as experience*. The Berkeley Publishing Group.

Dombrovskis, P. (2017). *Journeys into the wild: The photography of Peter Dombrovskis*. National Library of Australia.

Dunn, J. (2005). Practising the art of forensic assessment. *Drama Queensland Says*, 28(2), 2–5.

Eisner, E. (1985). Aesthetic modes of knowing. In *Learning and teaching the ways of knowing* (pp. 23–25). University of Chicago Press.

Flanagan, R. (2007, May). Out of control: The tragedy of Tasmania's forests. *The Monthly*. https://www.themonthly.com.au/issue/2007/may/1348543148/richard-flanagan/out-control

Gablik, S. (2004). Beyond the disciplines: Art without boundaries. *The Hedgehog Review*, 6 (2), 61–71.

Greene, M. (1978). *Landscapes of learning*. Teachers College Press.

Greene, M. (1999). *Variations on a blue guitar: The Lincoln Center Institute lectures on aesthetic education*. Teachers College Press.

Greene, M. (2007). Imagination and the healing arts. *Maxine Green Institute*. https://maxinegreene.org/uploads/library/imagination_ha.pdf

Greene, M. (2009). Obama, the schools, imagination, and convergence. *Harvard Educational Review*, 79(2), 396–398.

Griffin, S. (1996). To love the marigold: The politics of imagination. *Whole Earth Review*, 89, 61–67.

Hennessey, B., & Amabile, T. M. (2010). Creativity. *Annual Review of Psychology*, 61, 569–598.

Jacobs, R. (2009). Measuring aesthetic development: A national dialogue. *Journal of Artistic and Creative Education*, 3(1), 75–96.

Jefferson, M., & Anderson, M. (2017). *Transforming schools: Creativity, critical reflection, communication, collaboration*. Bloomsbury Publishing.

Kiernan, K. (2017). The original Lake Pedder, southwest Tasmania: Origin, age and evolution of an Australian nature conservation icon. *Geoheritage*, 11(2), 271–289.

Marks, M. (2017). Exploring paradigm shifts in environmental art to find a definition. In D. Curtis, (Ed.), *Building sustainability with the arts: Proceedings of the 2nd national ecoarts Australis conference* (pp.19–34). Cambridge Scholars Publishing.

Martin-Smith, A. (2005). Setting the stage for dialogue: Aesthetics in drama and theatre education. *The Journal of Aesthetic Education*, 39(4), 3–11.

Miles, M. (2014). *Eco-aesthetics: Art, literature and architecture in a period of climate change*. A&C Black.

Milne, C. (2017). *An activist life*. University of Queensland Press.

Moon, S., Rose, S., Black, A., Black, J., Hwang, Y., Lynn, L., & Memoli, J. (2013) Releasing the social imagination: Art, the aesthetic experience, and citizenship in education. *Creative Education*, 4(3), 223–233.

Press, C. (2018). *Rise & Resist: How to change the world*. Melbourne University Press.

Rankin, J. (2019, November 28). Our house is on fire: EU parliament declares climate emergency. *The Guardian*. https://www.theguardian.com/world/2019/nov/28/eu-parliament-declares-climate-emergency

Ross, M., Randor, H., Mitchell, S., & Bierton, C. (1993). *Assessing achievement in the arts*. Open University Press.

Steffan, A. (2014, April 28). The politics of optimism. *Alex Steffen*. http://www.alexsteffen.com/the_politics_of_optimism

Wallen, R. (2012). Ecological art: A call for visionary intervention in a time of crisis. *Leonardo*, 45(3), 234–242.

PART 3
Learning nature–culture

10

BEING EFFECTIVE

Social ecological understanding in action

Cathy McGowan with David Wright

Cathy McGowan, at the time of this interview, was the federal Parliamentary member for the regional seat of Indi. She was elected in 2013 and retained the seat in 2016, with an increased vote. She did not contest the 2019 election, but acted as mentor to her Independent successor, Helen Haines. Succession of this kind is very rare in Australian electorates, especially those in traditionally conservative rural areas. Cathy is a graduate of Western Sydney University, where she completed a Master's degree, built around study in systems agriculture and social ecology. This learning, she says, was central to her work as an elected member of the Australian Parliament.

> Well, I'm absolutely convinced, that for me it is all about experiential learning. And this is what I learned studying systems agriculture and social ecology at Hawkesbury, Western Sydney. When I first went there I thought there was a world out there and it assessed you and you got marks. I thought that was what university was. Well, Hawkesbury wasn't like that. They said to me, "you've got to learn to observe yourself." They said "there are processes of learning: ways in which you can see yourself in relationship to the world, and once you get that perspective you can start to think about your effectiveness." And they added, "Once you learn how to be effective you can also learn that there is not just one way to be effective." So, I learned about being effective during the two years of my Master's degree. I learnt what effective was. I learned by doing the work upfront. I learned that by knowing what I wanted to achieve, I was much more likely to achieve things.

I met with Cathy in her Parliament House office early in 2019. The intimidating "importance" of the house is difficult to put to one side. Cathy knew this and we relaxed into conversation through three artworks on the wall of her office. Through these she drew my attention to her electorate, her "place." Indi: 30,000 square kilometres of fertile valleys, of midsized regional towns, a slow moving river to the north, rising mountains to the south and east, an expansive river catchment – a social-ecological system that has been home to her family for many generations.

> Yes, I've got 9 sisters and 3 brothers and 5 of us still live in the area, so the community is really strong. My mum's family have been there since the gold rushes, so there's a lot of family history and a lot of extended family. That's where my roots are… that's where the family stories are. And that's where I grew up and went to school.

I recalled a conversation in 1996, when we spoke of problems facing women in rural Australia. I remembered talk of issues facing women. Issues of farm ownership, finance, family and inheritance: "*No matter how much she loves it, the daughter never gets the farm.*" Cathy spoke then, not as a student but as a nascent leader of women in Australian agriculture.

> I did a lot of work with women in agriculture. And we set up organisations and systems and did the lobbying, and this was very much because agriculture was dominated by blokes and we were fighting to be included. It was a real battle to be included, to actually get the men to make space so we could be heard. There was a lot of strategy about how we would do that. And we would get beaten back, then regroup and have another go. We learned a lot about how systems don't like changing… but also how you can change them.

It is not easy for an independent to be elected to Parliament, even less so for a woman in a rural electorate. Scepticism needs to be overcome and widespread support earned. Early in our discussion Cathy made it clear that representatives need to respond to their community. "*Community,*" she argued, "*is a process. It is an action that is learned.*" It is a feedback system, and the role of the representative is to be consciously active in that feedback process: to listen, learn and become immersed in the emergence of knowing.

> The credo I bring to my job in Parliament is "be the change you want to see." That's experiential learning. If you understand your own environment, the only person who can change it is you. So, it is not a matter of trying to change the world. What it amounts to is "get your act together and be the change you want to see." That is the way I approach this work: 100%. The other saying that I like is, "if it's going to be, it's up to me." Again, experiential learning.

But what does this actually mean?

> Here in this office I'm an independent Member of Parliament. I am the one vote between the opposition side and the government. Together, with the opposition, we can vote the government down. I'm acutely aware of the power and the responsibilities of the job, and the opportunities to bring changes that I would like to see to Parliament. It's no small thing.

So, how can you be the change you want to see?

> I've got to reflect upon who I am and what I do in my environment. This means understanding that I'm in a system and my being within that system has impact and impact creates change. And being strategic enables me to have greater impact. I'm very conscious of this in my work here.

How do you see yourself, and what is the change you want to bring about?

> I'm a country woman who is currently and utterly committed to community, not just the community I come from, but building community and making community in which other people belong. I do it in Parliament, and I do it in my home region. In this place [Parliament] we work to make a "belonging space," and not just for staff. Every sitting week we have up to four community people come and work in the office as volunteers. We have a process of inducting them. The instructions are simple. The culture of the place is respect, and the values of the office are community and respect and belonging. Everyone belongs here. And then we have a conversation around being here, in Parliament. We tell the volunteers that they don't have to pretend to be anyone or anything. We tell them that while they are here they need only be themselves: their best selves. The other thing we talk about is leaving all preconceptions at the door. There is nothing here that is what you assume it to be. Open yourself to it. Let your experience be the teacher.
>
> As the week unfolds we have a briefing each morning and afternoon, a formal reflection upon the day: "What have you done, what have you learned, what sense have you made of it." And, that's what I learned at Hawkesbury: reflective learning. There's no "right" about the world you're in, it's just the way it's unfolding or emerging. So, out of choice, I try to create community for all of us here: for all those people who come and share this place with me – and it works. We've had more than 180 volunteers from Indi up here since I started in Parliament. We've had people from all walks of life come up and get a taste of political representation. That's being the change you want to see. I want to build community here based on who I am, a country woman who is warm and hospitable and respectful of people. And the more I walk the talk, the more it creates belonging, warmth and liveliness in this corner of the building.

So, you build a bridge between your electorate and Parliament. And people back in Indi feel they are in connection with you, as a community facilitator?

> That's the programme: it's called "Community to Canberra." This week we've had local government people up. They've come up and we've organised appointments with everybody, and we've had a bit of a debriefing about how is this going to work at home. As a result of this they've created a two-day workshop on how they are going to work with Canberra next time. I haven't done it, they have, but I've created the space for it to occur. And that makes my community, where I live and what I have a commitment to, a better place to be. It's about me being the change I want to see.

This is an alternative to the conventional party system, where you have a party structure and party operatives and you need to work through them, rather than the whole community.

> Some of my colleagues here ask me, how can you do this peer-to-peer volunteer stuff. And I tell them, but I know they are never going to be able to do it because of the party- based systems they work through. They can't have their community come to Canberra; it has to be party members. And in a party there's rules – who's in and who's out – but as an independent I don't have the in and out problem. And the other thing about me is that because I'm committed to being the change I want to see, my emphasis is not on me making other people change. This means I can afford to give people opportunity. And this makes me a teacher of experiential learning. The focus becomes, how can we have people's needs met? And in providing that, people do the work that enables me to represent them (not some predetermined agenda). For example, one group that came here to do young peoples' extracurricular learning spent four days here. I introduced them to the Minister for Education, they had appointments with several Ministers, and they went from being sceptical about politics to thinking, "Oh my god, these people actually listen." And when we debriefed we said, why don't we invite them to Indi. And the Minister came to Wodonga, and was hosted by this group of people. As a consequence, the volunteers did research for me, drafted speeches for me, and helped me represent them in a way I never could have done if I was in a party. And they did it because they could see what information I needed to be their representative, and they could see that they could help me do a much better job of talking about what they were interested in – young people in education – because they had the knowledge and the enthusiasm and an increasing sense of purpose and possibility.

What to you is "effective"?

> Initially, when I started here, I didn't know. Initially I had no idea. At a very basic level I wanted to survive and come out the other side with my integrity intact with my community. Now my measure of "effective" is related to me being my best self: working in a respectful way in governance. It is about my imagining, my thinking that if the world could realise its potential, how might

it be. I remember the work I did with Judy Pinn [former Social Ecology staff member] in the unit Imagination in Action. She challenged us to think about this, and then added, "let's go and make it that way." So, I share my vision with other people and I get them involved. I think a lot about the best ways to get people involved, to give them an experience of belonging, of community, of being accepted and not being judged. Of getting everything interconnected. And if you want to change something, sometimes it's enough to change this fundamental understanding: to help people appreciate that we are all a part of this system and everything we do as individuals contributes to ongoing processes of change.

One of the things I remember from when you were a student is the passion you and other women had for regional Australia. You created networks throughout the countryside. You were constantly active out there. And beneath this was a strong sense of connection to the country: the place we in the city so often overlook or take for granted.

I remember one of you guys did a workshop, I think it was John Cameron [former Social Ecology staff member]. He asked us to us draw what we saw, not what we thought we saw. And it was just so challenging. But the process of doing that helped me understand the construct of land and how we construct what we think we know. That helped me to deconstruct how I understand and experience the land. This had a lot to do with input from Social Ecology. Systems Agriculture was much more pragmatic, much more focussed on achieving things, whereas in Social Ecology we learned the skill of imagining into action. I remember another one of the Social Ecology guys doing workshops on letting your subconscious talk. Your conscious mind is always running you, but paying attention to your subconscious enables a lot of different learning. It can give you greater knowledge about yourself and greater confidence to take action. Stuart Hill [Emeritus Professor of Social Ecology] has an activity he calls "The great lie." It's a simple exercise in injecting imagination into action. "The way you want the world to be; that is the great lie." To imagine this and then imagine the way you are going to make it happen: that is the game I play here, in Parliament.

So, let's think of Parliament as a place that works for everybody. If so how do I make it a place that really works for my electorate, and really works for young people. I can provide space and let young people do what they want to do in it. I have no ownership over what they do, but what I do know is that I can give young people, who come up here as volunteers, the opportunity. They come here with all sorts of assumptions and I say, "just have the experience, let it happen, then we'll talk about it." Out of that comes insight, which feeds confidence and out of that comes understanding and affirmative action.

OK, so, there is community politics, which you have come from, and there is institutional politics. Can you make the institution a community, for you?

> Yes. Yes. I do believe it's possible and I am so keen to talk about it. Last night I went to a meeting of the Parliamentary Friends of India. About 30 of us were invited to the Indian High Commission for dinner. There was genuine warmth there. We gathered not as members of political parties, but as representatives who were interested in India. It was not a political gathering, but it has a lot to do with the job, with the work we do as parliamentarians. What made it particularly useful, in terms of community, was the strategic work we, as politicians, did over the meal. So, I was able to come into work today having done serious work last night in a "family" environment. A dinner where you chat about what to do with the badly behaving uncle, the petulant aunt, the brother. A group of us were able to sort out a problem and the work is done as a community.
>
> I'm on another committee on constitutional recognition for indigenous people. I've just been to the north west of Western Australia, and the chair of the committee is absolutely committed to having our committee work like a community. He puts an agenda together, he talks with people, he's an introvert, and he focuses on the set up, and I get to do my extrovert stuff and make it work. It is the calibre of the people that stands out. When we were up north, Senator Pat Dodson, one of the traditional owners of the area, asked "Would you like me to show you around." He spent three and a half hours showing me around his home country of Broome, explaining the hopes and needs of the traditional owners. So, you find like-minded people and you work together and there is generosity of spirit. But it is a high-level skill. It starts from willingness. If you've got willingness and a shared purpose you can have great interaction here.

Let's talk more about the gender issue. You identify very clearly with women in rural areas and apparently your success in Indi had a lot to do with your ability to build connections between women in regional Australia.

> After I finished at Western Sydney I did the Australian Rural Leadership course. It was all about leadership development, and I remember one part of the programme. There were about 30 of us in it: 26 men and 4 women. The guys were all successful farmers. They pulled me aside one night and said they wanted to take me out to dinner, and over dinner they helped me understand that my effectiveness in that group was lessened by the way my strong feminism claimed the space. I took enormous umbrage at this. Men trying to tell me that I was too strong as a feminist. But over the 18 months of the course I experimented with other ways of being female in a male dominated world, ways that weren't so confrontational. By the time I had finished the course – practicing leadership and learning from and looking at other women leaders and the way they practised the

art of leadership – I had the opportunity to actually teach a course to women about leadership. This is where I really learned that when it comes to leadership styles it's horses for courses and there's a million ways of doing it. I learned that if you know where you want to go, and start with that in mind, it's possible to work backwards. "Beginning with the end in mind" has become a favourite saying. So that was useful. Since that time I have spent over 20 years running courses for women about leadership.

When I decided to stand for Parliament, I made a decision to proactively be the change I wanted to see. I decided I wasn't going to make gender my issue, and I was always going to be a proud woman – a proud rural woman. I was going to be true to who I am, and I was going to be bringing many, many women with me, because that was my network. I am keen to be seen to be an effective woman who can deliver. To do that I realised I needed to get a community conversation happening, that says "Cathy's really good value, she delivers." That was what I needed. And it had to be third party endorsement.

So, as we approached the 2013 election it became a matter of working with women to get the conversation happening. And they bought their husbands with them. Most of them wanted a change [in their local MP]. It helped that most of my competition [in the election] were women. Of the 12 candidates, the 5 leading ones were all women. We didn't have to do gender in an obvious way. I made sure that when we went to places I would always have my women's network there. It meant that we always had lots of women turn up, and they would bring the men with them. It was subtler than saying I'm a feminist and I'm a woman candidate. I wanted my way of operating to be women friendly, and not divisive.

You didn't have to be "the hero."

No, I had to be efficient, I had to be effective, I had to turn up, I had to be warm, I had to be engaging, and I had to do that in a feminine way. There's a line there somewhere. Before, I hadn't actually appreciated that there was a spectrum across which these things work. And now, in Parliament, I'm not so strident about my feminism. But my resolve hasn't changed at all. I understand how patriarchy rules the world, and rules Parliament, and I'm an operator in it and I'm always there as a good woman, standing up for other women. My approach in Parliament, and in my community work, comes from imagining – imagining this world into becoming a better place.

Tell me more about what Indi means to you.

When I came back to Indi after travelling and studying, I was able to find work in Indi – mostly community planning – so I got to know the region well. I got to know the people and the institutions. I gained a professional working knowledge of the region. I've written reports, I've done studies, I've lived here.

I know the people and I know their spirit. And when I was growing up, and even more so now, I've always been struck by the lack of advantage in my community. It always seemed other communities had more than we had. I often thought it was because of geography. Parts of the electorate are isolated. Lots of the constituency are farmers and some are poor. Unlike some in Parliament, I didn't have to go overseas to discover poverty. I found it in my backyard.

But the most immediate thing that got me elected was young people saying things like, "The train doesn't work … it's a crap train… mobile phones don't work … nobody listens to us … and no one's thinking about climate change." I knew it was true. So, there was that sense of the need to step up. Someone's got to do something. And it was these young people saying to me "Well, if you do it we'll get behind you."

The last few weeks in Parliament we have been debating tax. I did a survey of my electorate and asked them what their biggest issue was, and it was taxation. I was really surprised. Now, my electorate is traditionally conservative, but they just don't like the inequality that is in this legislation. Then we did more research about education, and discovered only 13.9% of my electorate have degrees and the Australian average is 22%. There was a pervasive sense within my community of being underprivileged and forgotten, a sense that things aren't as good as they could be. I've probably known it in a lot of ways, but now I've actually got the statistics, and I have this huge sense of responsibility. My question is always, how do we build community so the community can be the driver for the change they need? And how can I work up here, in Parliament, and help that change to happen?

For too long people in Indi have been hungover from the hardness of their lives. Nothing really positive had happened for so long. We were ignored and underdeveloped, but we are a really strong traditional country and regional community that is independent and resilient and still holding on. That is a really solid base to build on to make Indi the place we want it to be. I have a great sense of responsibility and a great sense of opportunity. And the more courageous I can be about offering ideas and supporting others, it might flow on to others. We are doing exciting work around environmental issues, community and renewable energies – really amazing stuff.

I'm an unusual independent in this business of politics. Most independents have been men, and they have a tradition of picking fights and can be righteous about their positions. I hope I can be the opposite and not righteous. My plan is to go under the radar. Maybe that's why I've been effective.

And your social ecological understanding has helped?

My social ecological understanding has helped.

11

TRANSFORMATIVE LEARNING THROUGH MĀORI MIGRATION TO AUSTRALIA

Roseanna Henare-Solomona

TE HURIHANGA AKORANGA MAI I TE HUARAHI MĀORI KI POIHAKENA

Prologue

Ko Tuhipa te Maunga, ko Tereawatea te Awa, ko Ngati Te Ara, ko Ngati Kopaki, ko Ngati Kahu o Torongare ki Waiomio, ko Tekau i Mua, ko Ngati Te Tarawa, ko Te Orewai me nga Hapu, ko Ngati Hine te Iwi, ko Ngapuhi Nui Tonu ahau. Ko Ititahi ahau, ara ko Roseanna Henare-Solomona toku ingoa...

Tena koutou e nga hunga tapu o tenei whenua, ki nga hau kainga ko te Darug, Gadigal me te Eora. Nga mihi nui kia koutou nga hapu mo tou aroha ki toku whanau e noho mai nei ki runga i tou whenua tapu. Tenei te tuku whakawhetai mo tou manaakitanga me tou matauranga mo te kaitiakitanga mana whenua. nga mihi, nga mihi, nga mihi...

In translation: I was born and raised in the beautiful Bay of Islands, north of Auckland in Aotearoa, New Zealand. My home is located near a geographical mountain or landmark known to us as Tuhipa. The river that provides the life force to our family is called Tereawatea. My genealogy connects me to various family groups who are descendants of a matriarch ancestor, Hineamaru. Through Hineamaru, I am directly linked to the ancient Ngapuhi ancestor Rahiri. I am Ititahi. I am also Dr Roseanna Henare Solomona.

I must acknowledge the traditional custodians of this land, in particular, the Darug and Gadigal people of the Eora Nation, whose ancestral lands my family and I live on. We are grateful for the opportunities we have enjoyed in your country. And as I share my thoughts within this learning space, I do so having great respect for your generosity, strong spirit and the knowledge embedded within the First Nation Custodianship of Country.

Introduction

Our stories connect us to the past. They create the present and also help to inform our future. My aim in this chapter is to highlight the importance of identity as an integral stimulus for learning. I will draw special attention to the value of empowering learners to embrace and build upon their own knowledge found in story, as a legitimate way of naming the world and their place in it. Lived experience, shared knowledge and stories handed down from past and present repositories offer understanding and knowledge that can be used to navigate new systems and changing environments. This is both liberating and transformational for the learner. It is also empowering for those with whom this experience, knowledge and understanding is shared, highlighting the value of reciprocal and collaborative learning. Themes such as culture, spirituality, relationships, place and identity enable courageous conversations about embracing personal knowledge to enhance new understandings. This information is likely to be of particular relevance and interest to the migrant population who relocate to Australia from various parts of the world. Such stories juxtapose the beauty of new beginnings in the face of challenges experienced while navigating two very different worlds: the old home and the new. Migrant learners often must negotiate the unfamiliar and, at times, cold and unfriendly terrain of the mainstream education system, with its competing ideologies that challenge personal understandings and sensibilities. With this also comes a sadness from becoming aware that Australia's First Nation people are often not visible and have limited input into the fabric of the migrant's new world and society.

Creative writing can help to deliver a powerful message, and so I have written this chapter in a variety of ways, consistent with the era, space and voices used to convey the messages shared. Furthermore, to stimulate honest and, in some parts, profound and provocative conversation, a social ecology (SE) approach is used as a framework to enable the most effective use of the stories, teachings and knowledge found in this work. Hill (2011) refers to this approach as a transdisciplinary metafield that is informed by various disciplines. He describes the methodology and structure of this learning system as one that can promote personal understanding, a centrality of relationships, and the importance of considered reflection in the construction of sustainable knowledge. As a learner, and particularly as a Māori woman, SE has been an enabling platform to incorporate my own traditional knowledge and lived experience alongside the dominant Western paradigm in which I now live, learn and work. It has also provided a lens to view the world and beyond, through my own eyes, without having to change who I am and what I believe, regardless of the power imbalance in our academic and social systems. The SE model can gently guide new learners through self-directed study, stimulating reflection, promoting questions and facilitating heartfelt discussion to enable personal transformative learning opportunities. For those who traverse worldviews beyond the Western paradigm, this process is critical in understanding how to navigate both spaces effectively (Henare-Solomona, 2012).

Social ecology model

The social ecology model presents a four-sphere approach of interconnected fields that can help individuals to navigate various pathways, creating patterns, forming relationships and building stories with people and place. By positioning oneself in the centre of the four domains – the personal, the social, the environmental and the spiritual – and then reflecting on time spent in each of these spaces, and participating in the relationships between them, one can begin to see how our lived experience helps us to understand the lessons learned over time and the knowledge gained therein. Furthermore, as we reflect on the past, present and future in each of these learning spaces, we see truths and realities about who we are, and the cultural norms, beliefs and values we live by. More importantly, one can learn about how we are connected to people and place. Lived experience, and the understandings that we can gain from this, helps us to know who we are personally, socially and spiritually. It can also enable us to develop and nurture our worldview, as we reflect on such questions as: Who am I? Where do I come from? Where am I going?

This work is also a reminder to those in positions of power (including educators) to consider the lived experiences and knowledge brought into the learning space by students. Cranton (2016) argues that educators can support learner empowerment in diverse ways, most of which involve small, ordinary, everyday interactions within the teaching and learning environment. She proposes that we need to become more conscious of power relations, including those that exist in the types of interactions we normally associate with democratic practice. In this context, teachers must recognise, value and embrace the existing knowledge students bring with them into the learning space. In doing so they can promote inclusion and empowered learning, which Cranton (2016) claims is both a goal of and condition for transformative learning. The empowered learner is then able to fully and freely engage in critical reflection, participate in meaningful discourse and effectively act on revised perspectives. This is a reminder that transformation happens when we begin to embrace our own knowledge and are empowered to be at the forefront of our individual learning. This means one must be courageous and share our tacit knowledge, even when we are the minority, or these storylines differ to the dominant group. Needless to say, this process can be both confronting and liberating. What is important here is to know that enabling learners to incorporate knowledge from their own experiences is a major step towards achieving transformative learning.

In support of my proposition, and to incorporate my own knowledge when possible, I have adapted the SE model to reflect my worldview and demonstrate how a Māori-enriched approach can be useful in this context. For example, the general SE map (central to Western ideology) usually starts with the personal, progresses to the social, then the environmental and ends at the spiritual. My SE map is quite the opposite, in that it begins at the spiritual, moves across to environmental, social and then ends at the personal. Although these two worldviews appear to be opposite, the flexibility of the SE model allows us to recognise common themes, intersections and various alignments and associations, extending the learning opportunities beyond anticipation.

Wairua – Spiritual

> **i te timatanga ko te kore**
> ko te po, na te po, ka puta ko te kukune, ko te pupuke, ko te hihiri, ko te mahara, ko te manako, ka puta i te whei ao, ki te ao marama e, tihei mauri ora…

In the beginning

In the beginning there was a void, within the void there was night. From within the night, seeds were cultivated. It was here that movement began – the stretching. There, the shoots enlarged and swelled. Then there was pure energy. Then there was the subconsciousness. Then the desire to know, Movement from darkness to light, From conception to birth, From the learning to knowing, I sneeze and there is life…

"Wairua" is a Māori concept that can be translated to mean spirit and spirituality. It is also one of the most difficult concepts to explain, as exact meanings and definitions can be as diverse as the wider Māori community itself. Nevertheless, as a starting point, I draw on the following explanation from Valentine, Tassell-Matamua and Flett (2017) who claim that for many Māori, wairua is best considered as a delicate *taonga* (treasure) imbued with *tapu* (spiritual restrictions), which invoke certain boundaries and constraints in its use and understanding. Tohunga (Māori specialists), the late Māori Marsden and Dr Rose Pere, regard wairua as being a fundamental part of our existence. Marsden described it as the ultimate reality for Māori, the source of existent being and life (Marsden & Royal 2003). Pere (1982) claims wairua is a natural phenomenon that has both physical and spiritual implications. I grew up in a very traditional way, under the mentorship of elders, and I was encouraged to learn about both the physical and spiritual aspects of our belief system. This included practising daily karakia (prayer), being mindful and respecting tapu (the sacred) and drawing upon the power of wairua (spirituality) to guide all I said and did and to enable me to live a happy and productive life as part of my whanau (family) and hapu (clan). Wairua can also be developed through such concepts and practices as manaaki (giving support, care and generosity to others) and sharing aroha (concern, compassion and love for people, our world and every living thing therein, including animals, fish, plants). This philosophical belief system has served our people well, and today it provides a constant reminder to us all that we are interconnected and interdependent on each other for survival.

Taio – Environment

he whakapapa - he kaitiakitanga

ko te whakapapa tēnei, mō ngā taonga tuku iho a io matua kore
kā moe a papatuanuku ki a ranginui, kā puta ko tānemahuta, ko tangaroa
ko tāwhirimatea, ko tūmatauenga, ko haumie-tiketike, ko rongomātāne
ko ēnei ngā taonga tuku iho ō rātou mā, ko mātou ngā kaitiaki mō ēnei taonga.

Our genealogical link to the environment

This genealogy recites for us our divine inheritance. Through the union of Earth Mother and Sky Father their children were born. The land and sea, the weather and the conflicts between the elements, the forests and the birds, the animals and plants. These offspring were entrusted to our care; treasures, given to us from the past, for our benefit to live and to manage for generations to come.

The environment is integral to our Māori identity and culture as we connect ourselves through whakapapa (our genealogical storylines) to the various natural elements. These narratives also link us to the great creator, which reminds us that all things are a part of an interconnected whole. We connect to mountains, landscapes, rivers, and to all the natural elements found in the environment through our whanau and hapu. Māori believe that the environment provides for our wellbeing and, in that context, we must recognise, along with this privilege, our obligation and responsibility to manage these resources sustainably for future generations. This commitment is conveyed through kaitiakitanga (the practice of guardianship and environmental management). We know our existence depends on a healthy environment, and so caring for and managing the land, forest, waterways, animals and the planet as a whole is necessary to ensure human survival now and into the future. Only when we remember to recognise ourselves as part of the environment can we begin to be effective in caring for our world.

Whakawhanaungatanga – Social

> **Watch, listen and learn**
>
> When I was very young my grandfather would come home from a walk around the farm with a hat full of eggs. I would ask "where did you get those from?" and he would smile and lift his brow. I knew he wouldn't say, so I began to watch him. All I saw was him watching something else – I was not quite sure what. He walked quietly, stood quietly and watched, then he came home. He did this every day, and each time I would ask, "what are you doing over there? and "what are you looking at?" No answer came, just a smile, and often a hat full of eggs. Then one day my excessive questioning became just a bit much for my Pop. He turned to me with a stern look on his face and said "kau e korero, titiro, whakarongo, ako" – "stop talking, watch, listen and learn."
>
> This small but significant teaching moment, helped me to understand the importance of observation, awareness, body language, insight and learning from experience. Through watching and learning, and listening and learning, it became much easier for me to comprehend the subjective nature of our world, and the complex learning systems we traverse day-to-day. Furthermore, through reflection, establishing connections, identifying patterns, building relationships, and constructing stories, the task of navigating our social system became simple and easy to manage. I now realise reflection and critical analysis happens when we stop talking and we watch, listen and learn.

Reflection and critical analysis was taught in diverse ways at home and in various social settings. I asked questions from night to day, and most times these long winded "how comes?" were not answered by my Mum and Pop (grandparents). They shook their heads, lifted brows, made hand gestures, and often told stories to suffice my hunger for knowing. I then asked my "how come" questions to aunties and uncles, and to whoever would listen and respond. Endless knowledge flowed from all directions, yet the challenge was always about how to decipher the information – what was important, relevant and useful and, of course, what to do with it. When I reflect on the lessons I received from those around me, my thoughts take me back to when I was a child growing up in a hapu (clan). I belong to numerous hapu (clans) linked through our whakapapa (genealogical blood lines). This system starts with our immediate family and then grows to a large community group of extended family clans. Under the guidance of elders and leaders, our hapu works together to ensure that everyone is looked after and supported. The children are nurtured and raised by all of the adults in the hapu. At the same time, our old people, when frail and in need of care, are often looked after by their children and grandchildren. This highlights how our social system is responsible to all of its people as a whole. Fundamental values and principles are taught daily to help the hapu understand how best to live as part of a collective group. The term tikanga (customs and protocols) provided the teachings necessary to help us navigate this complex system or clan groups, where lessons about spiritual values, social rules and ethical and moral expectations were specified, received and learned. These values, rules and moral expectations establish the order in which things are done for the betterment of both the individual and the group. It also involves a collective effort in addressing matters if these social expectations and guidelines are breached.

Learning to navigate our social system, with adults who care about you, helps young ones to understand, appreciate and love what it meant to be Māori. This was my experience. Also, this way of living, learning, sharing and caring for each other was a constant reminder to me that we were nurtured and loved by many. For example, as a very young child, my primary school teachers (who were actually my aunty and uncle through our hapu whakapapa) coached and mentored me beyond school lessons and, as a result, I was supported in participating in a range of sports activities well above my age group. Then, in my teenage years, various aunties guided me, and other young women my age, through the challenges associated with puberty, reproduction, boyfriends, babies and so on. Because these were serious matters, it was a time to be especially focused and attentive. We learned how to care for and respect our bodies, about the power of reproduction and vital strategies to keep ourselves safe. Our aunties would often say you can give life through "te whare tangata (the womb)," which is a blessing, "so make sure you do not violate this taonga (this gift) in any way."

Most times we just listened quietly, took the advice and, at best, followed their directions. Basically, we knew that if at any time we needed the support of an aunty or uncle, they were available and present for us. We were lucky growing up in this way: being educated, supported and loved by many.

Whakapapa – Personal

The me I am

I come from an ideal world, a place where I am safe to be me. Where everyone is rich, but no one has lots of money. Where people work hard, but no one is really

> employed. Where things are given and received, but no one is actually in business. Here we stick together forever and ever. In this ideal world, we live according to customs and our own ways of knowing. We belong to a whanau, and hapu, who in turn belong to Papatuanuku, our Earth Mother. Whakapapa (our genealogical storylines) informs our sense of being. It helps us to know who we are, where we come from, and where we are going. This whakapapa also reminds us that we are kaitiaki (caretakers of the land that we know intimately as Aotearoa). It is a gift entrusted to us by Iho Matua, the God of all living things. I come from an ideal world, a place where I am safe to be me.

I am Māori, and in my world a person's sense of belonging is often established and sustained through his or her capacity to connect to a family, community and geographical location. These connections also link us to an ancient ancestor, whose storylines we inherit as our own. Although much of our genealogical information has been passed on through chants, song and various forms of traditional dance, it is the conversations shared between people that has sustained our knowledge and identity across the generations. Today, Māori knowledge can also be found in the writings of our scholars. Ranginui Walker (1987) suggests that our Māoriness derives from being an active part of a family and kinship system. John Rangihau (1977) asserts that Māori identity is about learning traditions and customs to earn one's place in the tribal community. These learners, he claims, are taught under the mentorship of select leaders and elders. Educationalist and language expert Timoti Karetu (1990) considers that Māori identity is about one's upbringing and his or her knowledge and observance of the rites of passage in a traditional Māori way. This lays claim to the notion that Māori knowledge, including the ongoing development and sustainability of our identity, has been a work in progress, evolving as we grow in age and understanding. This validates the importance of experiential learning for individuals like me, who make sense of the world by doing, watching, listening and learning. This validates claims made by our scholars that identity is enabled through engaging, sharing, teaching and collaborating with others. Walker (1987) explains this learning process further, suggesting that the primary function of the whanau (family) was procreation and the nurturing of children. And although maternal and paternal bonds are clearly recognised, all adults, often referred to as uncles and aunts, are recognised by children as being in *loco parentis* (serving in the roles of a parent). Although adults in the whanau play a part in the nurture and education of the children, grandparents are often the main repositories of the storylines that are passed down through the generations. I was six months old when my grandparents decided to raise me as their own. I grew up and learned the old ways from them.

The power of transformation

There is a saying that "the decisions we make today, can change a million after that." This aphorism is often at the very heart of my thoughts whenever I ponder the changes I have encountered since migrating to Australia. As I reflect on the adjustments and adaptations to my spiritual, environmental, social and personal perspective, it is fair to say the transformations have been phenomenal. For example, my worldview was established and nurtured by the cultural values our grandparents taught me. They imparted simple philosophies often founded upon experiential learning, subjectivity and old traditional ways of knowing. Social systems provided the backdrop for personal and group development, and the link to the environment was based on a connection to identity and place. These characteristics were embodied in a spiritual narrative that framed my divine understanding of who I am, where I come from and where I am going. Migrating to Australia changed all of that and suddenly the world became bigger as I stepped outside our traditional domain and into the global arena. It was a defining moment, as personal transformations would soon lead to major changes in the way I saw the world and my place in it. Mezirow suggests there are ten phases adults go through when transformative learning occurs:

1. a disorienting dilemma;
2. self-examination with feelings of fear, anger, guilt or shame;
3. a critical assessment of assumptions;
4. recognition of one's discontent and the process of transformation are shared;
5. exploration of options for new roles, relationships and actions;
6. planning a course of action;
7. acquiring knowledge and skills for implementing one's plans;
8. provisional trying of new roles;
9. building competence and self-confidence in new roles and relationships;
10. a reintegration into one's life on the basis of conditions dictated by ones new perspective. *(Mezirow, 2000, p. 22)*

As I reflect on these phases and my transformative learning experience, it appears that at every stage there have been challenges and new-found knowledge and understanding. For example, the disorienting dilemma occurred soon after all the excitement of moving to a big city subsided and the novelty of the new experience had diminished. I realised the social systems, our traditional lands and Māori culture as I knew it were no longer close by. I often felt alone and longed for the homeland. Self-examination and feelings of guilt followed as the fear of losing my connection to people and place became a reality. Furthermore, the concern that my children would also be disconnected and eventually assimilate to this new home was unsettling. These are truths I am sure many migrants to this land have faced. As our resettlement journey progressed, we formed new relationships locally and created strategies to maintain connections to family in the homeland. Mezirow (2000) refers to this process as options for new roles, planning a course of action and developing skills to implement these plans.

As the years have progressed, so too has my ability to navigate both the old and new worlds with confidence and skill. In Australia, I live happily with my family in the suburbs. We now have nine precious grandchildren who were all born in Australia. I work hard to make a contribution for the betterment of this country in reciprocation for having received many opportunities to grow over the years. My links to the homeland are as strong today as they have ever been. I travel to Aotearoa regularly and help where I can with tribal and extended family business. On my last trip home, someone said "I don't believe you live in Australia, you still speak like us." When I explained I have been living in Sydney for 30 years, they were simply astounded. I suppose some things do stay the same, even when change is happening around us. I also like to think we can remain the same but be different too, if this is what we choose. As I ponder the last phase in Mezirow's transformative learning process, I am reminded that we reintegrate new ways of knowing into our life with conditions dictated by the new perspective we gain after reflecting on how we have progressed through all of the stages. These new ways of knowing are often built upon ongoing reflection, trial and error, courageous conversation, thoughtful decision making and experiential learning. They are also the foundation from where we can then move from one place, situation, position or stage to another without losing sight of what it is that is important to us. Migration to another country can be life-changing, particularly when it is coupled with immersion into another cultural world, one that is far removed from what we know or assume to be our normal. When I started writing this chapter it was in the hope I could share my own story about identity, culture, resettlement, change and transformation while living in and between two worlds. The challenges of assimilating and leaving the old world behind for opportunities of a better life in the new, or striving to find ways to hold on to what is precious in both the old (the whole of me) and new (my precious grandchildren) by combining the two. This is reason enough to ensure that the decisions we make today transform the way we live in the future.

Epilogue

In the earliest part of our resettlement to country, it seemed natural to move towards the social make-up of Australia's First Nations people. It appeared to me we had much more in common culturally. I think we still do. Unfortunately, this connection was not as quick or straightforward as I had hoped, and it took many years before I fully understood how to navigate this relationship. Today my best effort is to always acknowledge our First Nation hosts for their generosity and strong spirit, to identify my place as a visitor to this land and, where possible, to make this known to others who might listen. Recognising my place in country not only respects the host nation, but it also acknowledges the values and principles of my own people. Knowing my place in this land has been a powerful learning experience and has transformed the way I understand what it means to live well in country.

References

Cranton, P. (2016). *Understanding and promoting transformative learning* (3rd ed.). Stylus Publishing.

Henare-Solomona, R. (2012). *Whakaaro Rua; two ways of knowing; understanding how identity changes when Māori migrate across the Tasman to live in Australia* [Unpublished doctoral dissertation]. Western Sydney University.

Hill, S. B. (2011). Social ecology: An Australian perspective. In D. Wright, C. Camden-Pratt, & S. Hill (Eds.), *Social ecology: Applying ecological understanding to our lives and our planet* (pp. 17–30). Hawthorn.

Karetu, T. (1990). The clue to identity. *New Zealand Geographic*, 5, 112–112.

Marsden, M., & Royal, T. C. (Eds.). (2003). *The woven universe: Selected writings of Rev. Māori Marsden*. Te Wananga-o-Raukawa.

Mezirow, J. (2000). Learning to think like an adult: Core concepts of transformation theory. In J. Mezirow & Associates (Eds.), *Learning as transformation: Critical perspectives on a theory in progress* (pp. 3–33). Jossey-Bass.

Pere, R. (1982). *Ako: Concepts and learning in the Māori tradition*. Te Kohanga Reo National Trust Board.

Rangihau, J. (1977). Being Māori. In M. King (Ed.), *Te Ao Hurihuri: The World Moves On: Aspects of Māoritanga*. Hicks Smith.

Valentine, H., Tassell-Mataamua, N., & Flett, R. (2017). Whakairia ki runga: The many dimensions of wairua. *New Zealand Journal of Psychology*, 46(3), 64–71.

Walker, R. (1987). *Nga tau tohetohe; Years of anger*. Penguin Books.

12

PASSIONATE IMMERSIONS IN NATURE–CULTURES OF THE EVERYDAY

Jen Dollin

Developing the "art of noticing"

In the paradoxical era of the Anthropocene the need for transformative educational processes and socio-ecological learning is more urgent than ever. One response to this provocation is to call for a "passionate immersion in the lives of others" (van Dooren, Kirksey & Munster, 2016, p. 6) and to develop the "art of noticing" in order for humans to forge more ethical connections within this multi-species world (Tsing, 2010 p. 19). But how does one go about even commencing such an endeavour? How can one learn to "remake oneself" to become attentive to the social-ecological world we live in? Where would one start? Margaret Somerville (2010) writes that we need to think beyond remediation, revegetation and conservation to a new ecological consciousness and ethics that emphasises the importance of changing relationships to a place, as much as techno-scientific solutions. Likewise, Astrida Neimanis and colleagues conclude that "a deep understanding of environment cannot be divorced from human imagination, culture, and institutional and social practices" (2015, p. 80).

This chapter offers my engagement with this learning process and traces how I have worked through these questions in a social ecology framework deploying auto-ethnography as a critical creative/cultural writing methodology. Anthropologist Anna Tsing (2012) writes that "familiar places are the beginning of appreciation of multi-species interactions" (p. 142). And so, for this work I start where I am: deeply embedded in my local place, the Hawkesbury River in New South Wales (NSW) and (obscurely) with a short-finned freshwater aquarium eel (*Anguilla australia*) as my research partner. The ongoing entanglements with this specific eel and with freshwater eels in general – ubiquitous, uncharismatic and relatively invisible creatures – have triggered deep socio-ecological learning involving multiple dimensions: cognitive, affective, existential, empowering, and active empowerment and action (Sterling, 2011). It has generated a raft of new transdisciplinary self-learning that connects diverse fields such as fish biology, cultural theory, aquatic ecology, anthropology, environmental humanities and educational philosophy.

For me, this work has opened up new possibilities for thinking, writing and researching in a social ecology framework. It is not a story that has grand solutions for our current socio-ecological crises, but a personal journey of self-learning to be shared. Significantly, it is also not a story that I could have written five years ago: I simply did not have the words, thinking or discernment as I was deeply and unknowingly entrenched at that time in a positivist rationalist frame. This has very much been a case of writing my way forward into a new paradigm of passionate immersions. Please consider this my small but sincere offering on the remaking of relationships with nature–cultures of the everyday.

Rivers in the Anthropocene

Річка - Rychka – River, 1959
I can't remember when the river was clear
Mum said it used to be when she was young
Bilokur, Borysko, Celbulski and Sluvinski kids
Would sneak to the Hawkesbury River after school
European skins warmed by Australian sun
They learnt to swim in the dams at Camp
Then graduated to jumping off the pylons at Windsor Bridge
It was safe 'cause you could see the bottom
English, English, English they were told
But the Eastern Bloc stuck together
Polish, Hungarian, German, Ukrainian and Latvian
Rzeka, Folyó, Fluss, Річка, Upe
The other local kids NEVER acknowledged them
Not at school, on the bus, nor at the river
After all they were here first don't you know…

Despite acknowledgement that "water is strongly situated in local ecologies, histories and cultures," in Australia, attitudes about rivers and aquatic ecosystem management are deeply rooted in Western principles (Miller et al. 2014, p. 4). Notions of human supremacy based on a technical managerial approach rest predominantly on scientific knowledge. Urban and peri-urban river systems in Australia are impacted by urbanisation, industrialisation, vegetation clearance, agricultural irrigation and drainage, and this remains a sensitive issue for river management authorities (Pinto et al., 2013). The Hawkesbury-Nepean River system, an iconic but degraded waterway that threads its way around the edge of the Sydney basin, reflects this common story of the Anthropocene era. To add to this complexity, the damming of the river at Warragamba has altered the flows of the system. Inputs of treated sewerage water from the 16 wastewater treatment plants (WTP) situated in the catchment now contribute to these flows (Sydney Water, 2018). The river system is now subject to a swath of human-created pollutants such as antibiotics, endocrine-disrupting compounds and illicit drugs such as methamphetamine and cocaine that come through these WTPs with high levels of uncertainty about their long-term impacts (Reynolds et al., 2018)

The ecological consequences of these human interactions – at scale with the Hawkesbury-Nepean freshwater and estuarine wetlands, streams and rivers – while dynamic and changeable, have resulted in an overall decline in waterway health, increase in pollutants and invasive species, and loss of aquatic, vegetative and macro-invertebrate biodiversity (NSW DECC, 2010; 2011; NSW DPI, 2013). Federal and state governments, supported by the scientific community, have responded with the legislative classification of species and communities as "endangered," "threatened" or "vulnerable" to extinction, and the development of strategies focused on supporting recovery (Environment Protection and Biodiversity Conservation Act, 1998; NSW Threatened Species Conservation Act 1995; NSW Fisheries Management Act, 1994).

The Hawkesbury-Nepean catchment is awash with aquatic and ecological life that still cycles through its own rhythms, although now inextricably entangled with humans and their impacts. There are over 60 species of frogs and 50 species of fin fish, nine of which are introduced. Seven species of native frogs, a similar seven native fish species, and two dragonfly species are classified as a threatened species (NSW DPI, 2006). Fish passage and stream connectivity for migratory fish species, competition with introduced species, and the decline of water quality continue to be major threats for these aquatic ecosystems. In addition, whole communities of vegetative riparian ecosystems, including the array of creatures living within them, have been classified as "endangered ecological communities (EECs)" (NSW DPI, 2006).

Despite evidence of such ongoing ecological damage, economic rationalism persists as the basis for NSW natural resource management and decision making. The environment is regarded as a passive, empty landscape that is available for human intervention and activation. Successive failed management regimes instigated by a range of short-term programme reinforce the challenges. Wicked problems require long-term interdisciplinary research and "there is an urgent need to investigate the way different knowledges can be integrated with each other and practice" (Sofoulis, 2011, p. 45). Several years ago, I was obsessed with understanding "how to save this river," "what needs to be done" and "who needs to take action": local councils, state government departments, politicians, or all of these? This thinking is indicative of a classic positivist approach: something needs to be done, something should be fixed, there is a rational scientific solution, and then the river system and everything in it will be "better."

I can see now there are no ultimate answers, only changes in one way or another. I too am implicated in the river and its cycles, entangled by my own bodily functions and family memories. I am not an outsider; I am not a bystander. Rather, I am deeply immersed in this place and implicated in the health and wellbeing of our rivers. Personally, I try to tread lightly, participate in bush-care activities along the river, undertake river walks and pick up rubbish. It's not much, but it is what a working, exhausted middle-aged women can do in her spare time. Professionally, I sit on local advisory committees, seek and implement riparian revegetation grants, participate in "rivertalks" and community consultations, and write reports about systemic mismanagement and policy dysfunction.

Child of the riverlands – Why eels?

вугр – Eel, 1963
Apparently we grew up eating eels
Although Бабушка reckons she *never* touched them – нет
Mum said my Дед Constantine Yakobovych loved them though правда!
Because he was a farm boy from Hutin
(As if that explained everything)
I like to imagine him sitting soberly on a river bank with his line
But it's more likely he went there to drink vodka without questions
Maybe he imagined fir trees along the creek instead of gums
Or maybe he just wanted to forget the horror of Hitler
Luda remembers he would light an open fire ringed with stones
In the far corner of the backyard next to the chicken shed
He would roast the картшка they grew in the garden
And half smoke the eels on long sticks, then everyone would eat –
Everyone except Бабушка: she would stay in the kitchen
Far away from the vodka, smoked eels and memories of war.

Journal, May 2018

Translations

вугр – vuhor (eel)
Бабушка – Grandmother (Babushka)
Нет – no (nyet)
дедушка – Grandfather (Djedushka)
Правда – truth (pravda)
картшка – potatoes (kartoshka)[1]

When my patient long-suffering doctoral supervisor suggested I document my family and cultural stories of growing up on the Hawkesbury River as part of a legitimate research process I felt deeply challenged. I just wanted to write about and research watery river worlds, not about *me*. Research was something to be done objectively, at a distance, emotionally isolated and safe! "Why eels, why the river, why do you care, what is your connection here?" she kept asking. I kept resisting. It was only after, yet again, half-heartedly reading *Body/Landscape Journals* (Somerville, 1999), and *The Ethnographic I: A Methodological Novel about Autoethnography* (Ellis, 2004), that things made sense and fell into place. Drawing on Ghandi, I realised that I can only remake myself, not the world, and this is where I need to start (Edberg, 2019)! This auto-ethnographic practice has been invaluable in opening up a deeper level of socio-ecological thinking and has required a learning process that is individual, intimate, attentive and place-based (Ellis et al., 2010). I still undertake the outlined activities; however, I do this with new insights and a more sensitive understanding of ethics and the entanglements of my daily life with that of the river and the many creatures (including eels) that live within it.

It's taken me a long time swimming through these family stories and memories to understand my own place in these riverlands. I consider myself a child of the Hawkesbury, and you could say I was spawned in the reaches of the Upper Hawkesbury River. I have a strange affinity with eels and the rivers in which they dwell for most of their lives. This could be because I myself am a product of dual cultures and multiple landscapes. On my father's side, we have deep settler Anglican roots, stretching back to 1802, and to the wave of English free settler-invaders who grabbed land in Yarramundi and had it declared "mine!" On my mother's side, we came to the Hawkesbury via the massive post-WWII migrations. This branch came with baggage and expectations: an Eastern European socialist perspective of community and work ethic, combined with the pressures of having their first born in a new country – a child of upheaval and longing for lost places, language, culture and advancement. Growing up, we were acutely aware of our status as "New Australians," a term no longer in common use. The name was coined by Australia's first Minister for Immigration to promote assimilation. However, my father's family (basic, working class Anglo-Celtic in origin) would always say it with a condescending sniff.

Like eels, refugees in Australia are still not exactly loved, being viewed as an uncharismatic but tolerated species at best. River eels are universally found across Europe, North America and Asia Pacific. They are prime predators in their own watery environments, with a complicated life history. They are highly adaptable and comparatively long-lived: up to 50 years (NSW I&I, 2008). Australian river eels can grow up to 165 cm long and they eat a wide variety of foods, including small crustaceans, insects, fish and frogs (Jellyman, 1989, NSW I&I, 2008). River eels are catadromous fish, that is, they live in fresh and brackish water but migrate up to 3000 kilometres to breed somewhere in the deep Coral Sea. It is still unknown exactly where they breed and it is presumed that they die after breeding (Castonguay & Durif, 2016). Because of this migratory behaviour, no Anguillid species can be bred in captivity.

Globally, river eel populations across Europe, North America and Asia are in dramatic decline, with growing uncertainties about the long-term viability of the Pacific populations. Outside a concerned cadre of scientists and aquaculturists, the global decline of these species, with those in Europe being classified as "critically endangered," has raised nary a blimp on the conservationist's radar (Jacoby & Gollock, 2014). They have been in steady decline since the early 1980s in Europe and North America, where recruitment of glass eels has declined by 90%. Reasons for this include a mix of exploitation and trade, habitat decline, pollution, parasites, predation and disease, climate change and changing oceanic currents.

Unloved creatures and passionate immersions

"Have you ever felt an eel?" When I asked this question of a young ecologist at the University campus where I work, she visibly shuddered. She had spent her thesis field work in the dams and wetlands of our campus, thigh deep in waders netting, measuring, sexing, weighing, tagging and releasing almost 1000 eastern long-necked turtles (*Chelodina longicollis).* Half of the bycatch in this process were the only two species of Australian eels found on the east coast: the short finned *Anguilla australis* and long finned eel *Anguilla reinhardtii* (Ryan, Burgin & Wright, 2015).

> "Ugh!" she exclaimed, "Slippery!"
>
> Journal, November 2017

Time and time again I hear this standard response about these – to me – amazing creatures. It is worrying for me. It prompts wondering. How can we as humans come to care more deeply about those that are so different from ourselves they illicit distaste? What does this mean for those unloved creatures of the world? As Elspeth Probyn writes:

> Is it easier to care about terrestrial food than seafood? The land rather than the oceans? Farmers rather than fishers? Obviously we care for some species more than others simply on their good looks and good luck to be anthropomorphically cute. It's hard (though not impossible) to cuddle a fish.
>
> *(Probyn, 2015, p. 76)*

How can I learn to be affected by the world in my daily life in my daily places? As I noted earlier, one of my research partners in this work is an aquarium short-finned freshwater eel: *Anguilla australis*. This eel's life is a negotiated, captured one, an outcome of the ornamental fish trade and limited to a 6-foot-long glass aquarium. "Spence," as we call him, does not have free will in this exercise, and he participates unknowingly, exercising agency from an aquarium in interesting ways that I am only now, as co-participant in this research, beginning to comprehend. My research with Spence involves respectful observation, creating "thick descriptions," and engaging in "polite inquiry while visiting" (Haraway, 2015). Usually this involves sitting quietly in an armchair next to the tank, hypnotised by his sinuous movements, or being involved in the active, daily interactions of aquarium feeding and care. During my three years of engagement with Spence, he has become an integrated part of my life, and that of our office where the aquarium is housed and where feeding, tank cleaning and water aeration is carried out. The more I learn about eels, the more I am intrigued by them, the more I feel ethically compromised by Spence, and the more I care.

Journal extracts: Learning to be affected

Our whole office (well the three of us that work together) have become gradually immersed in the "arts of noticing" and are now fully entwined with Spence's life (Tsing, 2010, p 19). Spence was purchased as part of a living gallery aquarium installation, with a number of other fish, as well as some yabbies (crayfish), all being members of the freshwater aquatic species of the Hawkesbury River. Right from the start, we discovered that Spence was different. The morning of the gallery opening, the Waste Manager was called by a hysterical cleaner who had found a "snake" (actually Spence) on the floor next to the aquarium. We were amazed that Spence survived being picked up and returned to the aquarium, and that's how we discovered the term *catadromous*: which includes being able to survive on land and in water, both fresh and salt (Jacoby & Gollock, 2014). We call Spence a "he" because we named the eel after a local identity. However, young eels are sexually non-differentiated and we don't actually know what sex he is (Davey & Jellyman, 2005). Spence is more likely to be a "she," having been found in a freshwater dam behind the aquarium from where she was obtained, whereas male eels tend to be found in brackish estuary waters (Usui, 1991).

Three years later, and in a possible parallel of art imitating life, Spence as the supreme predator has outlived all his companions. The bass, catfish and yabbies have all died. We found them either suddenly expired and/or attacked by parasites or viruses, with nothing we could do to but watch helplessly as all the prescribed treatments failed. Fish death is a difficult thing to watch. Spence, at the time being smaller than his companions, would take baby eel-sized bites out of any fish that had become a corpse overnight. I can't remember how it started, but we questioned the appropriateness of the frozen blood worms that we were feeding him. He would gobble the still frozen worms so quickly we worried about the coldness in his eely digestive system. It was Helen, I think, who decided to try earthworms. It was horrible and yet rewarding, but each work day one of us would take a spade and the lime green cup, and we would go into the landscaped gardens of our historic grounds to dig up worms. Sometimes we gently squabbled over who would get to do it. In rain, heat or cold we knew that we had to feed Spence. We all came to know where the worms live, the sort of places they prefer in the garden and how they came to the surface in the rain. The smell of the leaf litter after rain is earthy and seems to be the most attractive place for fat, long worms. Spence doesn't "like" frozen blood worms anymore!

We know Spence can't see well, but he comes to the surface of his tank when we open the lid and use a finger to agitate the surface. We hand feed him by dropping the worms and watch them disappear in a sudden bite. Every now and then one of us will run a finger along his beautiful grey skin, as soft as velvet. I think we kid ourselves that he recognises our faces, however when visitors come he does tend to hide. Spence lounges in some favourite places in the aquarium, and will often bury himself completely in the gravel. We are researching how to install a suitable sandpit for him. There are periods of sinuous activity that are magnetic to watch. Sometimes we eat lunch next to the tank for the pure pleasure of watching her. Both Helen and I know that Spence could outlive us, and we worry about his longevity and care. As Spence is becoming larger, we sporadically discuss releasing him. It is a total conundrum for us. The dams near our workplace are full of eels. Would she survive if we release her? Would she be eaten by a larger predator? What would she eat? Could Spence make that incredible ocean migration and fulfil her biological destiny? Have we domesticated and ruined her? We are all so co-implicated in this little glass eel world that we don't have clear ethical answers anymore.

Nature–cultures and the everyday

Nature–cultures are found in contact zones which, in this multi-species world, humans are surrounded by and immersed in. For my work, these contact zones of the everyday have been found in my office and the waterways near where I live. The "contact zone" is a term coined by Mary Louise Pratt in her book *Imperial Eyes: Travel Writing and Transculturation* (1992), in which she describes disparate cultures meeting and trying to come to terms with each other. In *Where Species Meet*, Donna Haraway (2008) expands the concept to include ecological intersections and their meanings. Both interpretations emphasise that such contact zones are best understood as meeting places for interactions with others, often involving hugely uneven power relational hierarchies.

Although Bruno Latour (1993) used the phrase "nature–culture," it is Donna Haraway's (2003) conceptualisation in *The Companion Species Manifesto* that I am drawn to. Haraway contends that it is in the close contact zones where we can see how nature is not separate from our culture, but rather reveals an entangled, mutually co-constituted and co-generated socio-ecological reality. The idea of such "*nature–cultures*" is thus not a fixed representative point, but a locally situated, ever-evolving worlding or "becoming". Eben Kirksey and Stefan Helmreich (2010, p. 546) build on Haraway's concept of contact zones and outline multi-species studies "where lines separating nature from culture have broken down, where encounters between *Homo sapiens* and other beings generate mutual ecologies and coproduced niches." Anthropologist Anna Tsing's ethnographic account in *The Mushroom at the End of the World* (2015), about the matsutake mushroom, illuminates these multi-species interactions, describing how human and ecological worlds are mutually co-constituted in a tangle of messy assemblages. Tsing's work has been an inspiration for my research.

Through the learning process of auto-ethnographic writing and journaling I have explored the nature–cultures of my own family journeys and memories and the rhythms of office life centred around a small aquarium. I have examined how small details and events are enmeshed with larger ethical dilemmas and questions of connection to and understanding of local places. While initially this was very difficult – as it was emergent iterative process – the writing experience has illuminated my understanding of myself as a researcher and the power of transformative learning. It has been a profound exercise in experiencing the paradigm shift from positivism to post-positivism, and in becoming passionately immersed in (and even more grateful for) the everyday.

Note

1 The above translations are a mix of the Russian and Ukrainian words that we grew up with in my grandmother's home in Western Sydney. Babushka's mother was Russian and her father Ukrainian, and she can speak and/or understand six languages. She once told me, however, that she dreamed only in Russian because it is the language of her heart.

Acknowledgement

The philosophy of social ecology taught at Western Sydney University is a pervasive, transdisciplinary one that encourages its practitioners to move themselves in new directions of their own making. It brings together a lively metafield of disciplines that push and pull at each other, but always allows its scholars to question and move the field and discourse outwards (and inwards). The tradition, stemming from the then halcyon days of "systems agriculture" of the 1980s – known as the "Hawkesbury Experiment" – is one that values personal understanding, experiential and participatory learning, systems awareness and qualitative, creative methodologies (Sterling, 2011; Wright et al., 2011). Stuart Hill's social ecology taxonomy argues that an effective social ecologist requires understandings that bridge the individual/spiritual, social, ecological and advocacy domains (Wright et al., 2011). I'd like to thank David Wright and Stuart Hill for the opportunity to share my work and for welcoming me into the broad and exciting social ecology family.

References

Castonguay, M., & Durif, C. M. F. (2016). Understanding the decline in anguillid eels. *ICES Journal of Marine Science*, 73, 1–4.

Davey, A. J. H., & Jellyman, D. J. (2005). Sex determination in freshwater eels and management options for manipulation of sex. *Rev. Fish Biol. Fisheries*, 15, 37–52.

Edberg, H (2019, March 4). Gandhi's Top 10 fundamentals for changing the world. *The positivity blog*. https://www.positivityblog.com/gandhis-top-10-fundamentals-for-changing-the-world

Ellis, C. (2004). *The ethnographic I: A methodological novel about autoethnography*. AltaMira Press.

Ellis, C., Adams, T. E., & Bochner, A. P. (2010). Autoethnography: An overview. *Forum Qualitative Sozialforschun /Forum: Qualitative Social Research*, 12(1). http://dx.doi.org/10.17169/fqs-12.1.1589

Haraway, D. (2003). *Companion species manifesto: Dogs, people, and significant others*. Prickly Paradigm Press.

Haraway, D. (2008). *When species meet*. University of Minnesota Press.

Haraway, D. (2015). A curious practice. *Angelaki*, 20(2), 5–14.

Jacoby, D., & Gollock, M. (2014). Anguilla anguilla. *The IUCN Red List of Threatened Species 2014: e.T60344A45833138*. https://dx.doi.org/10.2305/IUCN.UK.2014-1.RLTS.T60344A45833138.en

Jellyman, D. J. (1989). Diet of two species of freshwater eel (Anguilla spp.) in Lake Pounui, New Zealand. *New Zealand Journal of Marine and Freshwater Research*, 23(1), 1–10.

Kirksey, S. E., & S. Helmreich. (2010). The emergence of multispecies ethnography. *Cultural Anthropology*, 25(4), 545–576.

Latour, B. (1993). *We have never been modern* (C. Porter, Trans.). Harvard University Press.

Miller, F., Bolitho, A., Jamieson, N., Catmur, C., Hurlimann, A., & Bowen, K. (2014). A plan to push limits? Investigating the ecologically sustainable development dimensions of Melbourne's central region sustainable water strategy. *Australian Geographer*, 45(1), 19–35.

Neimanis, A., Åsberg, C., & Hedrén, J. (2015). Four problems, four directions for environmental humanities: Toward critical posthumanities for the Anthropocene. *Ethics and the Environment*, 20(1), 67–97.

NSW Department of Environment, Climate Change and Water [NSW DECC]. (2010). *Lower Hawkesbury-Nepean River nutrient management strategy*. Department of Environment, Climate Change and Water. http://www.environment.nsw.gov.au/resources/water/10225hnnms.pdf

NSW Department of Environment, Climate Change and Water [NSW DECC]. (2011). *Hawkesbury–Nepean River environmental monitoring programme: final technical report*. Department of Environment and Climate Change. https://www.environment.nsw.gov.au/-/media/OEH/Corporate-Site/Documents/Water/Water-quality/hawkesbury-nepean-river-environmental-monitoring-program-final-technical-report-09112.pdf

NSW Department of Primary Industries [NSW DPI]. (2006). *Reducing the impact of road crossings on aquatic habitat in coastal waterways – Hawkesbury–Nepean, NSW. Report to the New South Wales Environmental Trust*. NSW Department of Primary Industries.

NSW Department of Primary Industries [NSW DPI]. (2013). *Hawkesbury–Nepean River recovery program: Final Report*. NSW Office of Water.

NSW Industry and Investment [NSW I&I]. (2008). *Status of Fisheries Resources in NSW, 2008/09: River Eels*. http://www.dpi.nsw.gov.au/__data/assets/pdf_file/0007/375928/River-Eel.pdf

Pinto, U., Maheshwari, B., & Ollerton, R. (2013). Analysis of long-term water quality for effective river health monitoring in peri-urban landscapes – a case study of the Hawkesbury-Nepean river system in NSW, Australia. *Environmental Monitoring and Assessment*, 185 (6), 4551–4569.

Pratt, M. (1992). *Imperial eyes: Travel writing and transculturation*. Routledge.

Probyn, E. (2015). Listening to fish: More than human politics of food. In P. Vannini (Ed.). *Non-representational methodologies: Revisioning research*. (pp. 72–88). Routledge.

Reynolds J. K., Russom B., & Ryan M. M. (2018) The presence of illicit compounds in the Hawkesbury–Nepean river system. In G. Vietz & I. Rutherfurd (Eds.), *Proceedings of the 9th Australian Stream Management Conference* (pp. 477–483). River Basin Management Society.

Ryan, M. M., Burgin, S., & Wright, I. (2015). Effects of wetland water source on a population of the Australian eastern long-necked turtle *Chelodina longicollis*. *Water Air Soil Pollution*, 226(12), 408.

Sofoulis, Z. (2011). *Cross-connections: Linking urban water managers with humanities, arts and social sciences researchers. Waterlines Report Series No 60*. National Water Commission.

Somerville, M. (1999). *Body/landscape journals*. Spinifex Press.

Somerville, M. (2010). A place pedagogy for "global contemporaneity". *Educational Philosophy and Theory*, 42, 326–344.

Sterling, S. (2011). Transformative learning and sustainability: Sketching the conceptual ground. *Learning and Teaching in Higher Education*, 5, 17–33.

Sydney Water (2018). *Sewage treatment system impact monitoring program. Interpretive report 2016–2017. Trends in WWTP nutrient loads and water quality of the Hawkesbury–Nepean River*. Sydney Water. http://www.sydneywater.com.au/web/groups/publicwebcontent/documents/document/zgrf/mtk5/~edisp/dd_199230.pdf

Tsing, A. (2010). Arts of inclusion, or, how to love a mushroom. *Manoa*, 2(22), 191–203.

Tsing, A. (2012). Unruly edges: Mushrooms as companion species. *Environmental Humanities* 1, 141–154.

Tsing, A. (2015). *The mushroom at the end of the world: On the possibility of life in capitalist ruins.* Princeton University Press.

Usui, A. (1991). *Eel culture* (2nd ed.). Blackwell Scientific Publications.

van Dooren, T., Kirksey, E., & Munnster, U. (2016). Multispecies studies cultivating arts of attentiveness. *Environmental Humanities*, 8(1), 1–23.

Wright, D., Camden-Pratt, C., & Hill, S. B. (Eds.). (2011). *Social ecology: Applying ecological understanding to our lives and our planet.* Hawthorn.

13

PLEASE EXPLAIN!

Brendon Stewart

Introduction

This chapter is about the "small arcs" of ethnically specific language communities braiding into the "larger circle" of an overarching language ecology. The chapter is arranged around fieldwork research carried out in the Sydney suburb of Auburn. Coincidently, the nature of this study is implied by the title of Nora Bateson's (2016) book, *Small Arcs of Larger Circles*. The chapter builds on the idea of a conversation that offers a transformative learning experience. Meaning is being made all the time as people engage in conversation, and conversations encourage communities to learn by way of negotiations (Stewart, 1999). Subheadings indicate what learning may be under discussion. The most recent demographic data for Auburn tells us that the community comprises a large and diverse immigrant population. The 2016 census indicates that in Auburn 4.6% of people had both parents born in Australia, and 84.6% of people had both parents born overseas.

Making meaning

A multicultural policy was instigated as an Australian Federal Government programme from the early 1970s. Since its integration into our laws and ethos, it has changed significantly. In the most recent incarnation, it has come to mean that migrants within mainstream Australia are free to, and should be encouraged to, express their cultural identity. It now refers to the notion that the people of Australia have multiple cultural and ethnic backgrounds. The people of modern Australia are immersed in the constant (and unavoidable) process of constructing their own cultural future. The more considered and informed this construction is, the more we exchange knowledge and engage with the transformative nature of this social learning, the greater will be our opportunities for social ecological endurance. The intention of my research is to provide a voice – as part of a wider social conversation that contributes to our social meaning-making – as we share and live in a multilingual community.

The focus of the research was an inquiry into how language communities other than English speakers can stay viable. For this to happen they cannot rely upon speakers of the language alone. There has to be a language ecology, embedded in a social ecology in which the person(s) speaking finds appropriate social and communal relationships. Although people capable of speaking a language are a necessary part of that system, they are not enough (Rhydwen, 1995).

Research participants are included in the chapter and their voices are identified in *italics*.

People tell stories to elaborate and help comprehend their world. Paul Carter (1992) in his book *Living in a New Country* says of immigrants that their story is often an autobiographical fiction in which poetic devices create metaphorical connections where more logical ones fail. The complexity of the contemporary cultures emerging here, on this large South Pacific island, are not determined alone by the biological and geological eccentricity of the landscape, but by some Antipodean reconstruction of the myths and fantasies from the various "old countries." According to Mezirow (1991), to reconstruct learning stories from the old countries involves transforming previously understood meaning schemes and meaning perspectives.

Acting on experience

For recently arrived non-English speaking migrants there may be times when their language becomes meaningless, when their words bear no relationship to familiar things. Language here will seem to have lost its fit. In these situations, the happenstances of social living may be the best means of getting on: pantomimic imitations of other people can bridge the gap. This happens day by day on the streets and in the schoolrooms. It is the transformative way individuals and groups act on their experiences of language use and learning. Tolerance and imagination are needed to embrace the cultural learning involved in understanding the metaphorical ways in which translation happens.

KOKSAL: *Going to picnics, this sort of developed later on, when people start buying own cars and things. As the time went on, people sort of start meeting at different places. Other thing that happened was we Turkish was giving names to small parks and places, in Turkish and meeting you there. Rather than using the Australian name, you know.*

We humans have evolved to imagine and then embed all living ecosystems into our broad-reaching cultures. Gregory Bateson (1972; 1979) argues that a social ecological epistemology must involve us understanding complex living systems that contain individuals, communities/societies and places and localities. Bateson understood that the study of ecology had to also be broad-reaching, bringing together the fields of biology, cognition, art, anthropology, psychology and information technology.

Language communities embody cognitive, anthropological and psychological systems of cultural complexity. An ecological epistemological approach as described by Bateson understands the nature of complex ecosystems as being in a state of connectedness, interdependency and change. Paradoxically, what maintains and sustains an ecosystem, including social ecosystems, are fluctuations and instabilities within this interdependency, constantly occurring at local and small arc levels throughout the system. Ethnically specific language communities maintain their viability by sharing their knowledge with other members of that language group, as well as spreading into the wider diverse community. Communal relationships emerge and knowledge is exchanged in a complex social ecosystem, such as in Auburn, through places of religious observance, schools, and through interactions with health professionals, the police, local government officers, shopkeepers, government institutions and a score of other connections. A less obvious, but essentially invaluable, connectedness and interdependency occurs in the broad reach of the arts and other aesthetic and imaginative fields of endeavour. Through film, television, theatre, sport, musical composition, performance and literature, societies share and communicate their knowledge and stories.

Communicating action

A wonderful example of this knowledge sharing is the monthly meetings of the Auburn Poets and Writers group in the Auburn municipal library. The group is multilingual and participants share, encourage and discuss their writing and work towards performance opportunities. As performance poets, they have participated in the Sydney Writers' Festival for more than a decade. Ivor Indyk (2008) says of this group that the many languages and different identities are central to the way their poems and stories emerge. The loss or limitation of language can be the most serious and debilitating consequence of migration, but nowhere here is this a tragic consequence. This poetry group exemplifies a process of symbiotic interchange that involves complex connections between people and their environments. It works by way of a cultural *poeisis* that generates a creative exchange involving improvisations of contact between people from diverse language and heritage communities. The poetry group communicates transformative action. Dorothy Makasa's poem *Beyond My African Consciousness* reminds the reader that she is a mother and a symbol of her "dark skinned continent" from whence she came (Zimbabwe). And so together we are grounded in an unfamiliar beyond.

> I am a mother, a member of the clan.
> A national symbolic representation of my dark
> Skinned continent.
> All necessary baggage,
> For therein lies the individual and collective pride.
> Grounding me in the unknowing of the beyond.
>
> *(Makasa, 2008, p. 51)*

Progressive development in our multicultural philosophy has been transformative for conventional teaching. As such, almost without exception, all schools in the local government area of Auburn have engaged with this new learning. This is especially so for the year 9 and 10 classes at the high school. The success of their endeavours, nonetheless, is specific to the school and the teachers. It is important to note that recently arrived migrant children are not the only disorientated people in a classroom. The teacher and other classmates, whose spoken language may only be English, will find the mix and flow of languages complicated. A good teacher will aim to create a learning environment that will enable transformation.

NESRIN: *I was thirteen when I came here from Turkey and start from year eight at Auburn Girls High School. I can't remember really how many girls exactly there, in my class. We had two different special English classes, yes. There weren't many Chinese then, mainly Lebanese, Turkish, Italians, Spanish, South American students. We were all put in the same class with special teachers. For about one year or so. We were mainly concentrated on learning to speak English, writing. We communicated whatever we learnt. We spoke with each other but mainly sign language at first.*

This is why our schools are so important. As public institutions they serve to underpin the process of socialisation for individuals and communities. This socialisation happens as children are exposed to the school's English language community, and curiously with the daily "clash" of the various tongues in the playgrounds, on the streets and in their homes. Although English is recognised as the primary language for most social transactions, it is important to note that Australia does not have an official language.

Language is emotional: the mother tongue, the whisper of memory and the transformative power of language learning, language sharing and language demands all put great pressure on our schools' teachers. The task is for the teachers and students to understand the transformative nature of learning by helping all expand their consciousness of a worldview that is diverse (Elias, 1997). This transformative learning is facilitated through consciously directed processes.

NESRIN: *I actually like meeting other people from other countries. You get to know the way they dress up and the way they eat. You go past their shops and you know, you see them, their families. I think it's very nice way of experiencing, finding out how they live and what they do and how they look. Now I think I can determine by looking at the person whose walking in the street. I can probably distinguish that they're from Spain or they're Lebanese or they're Turkish. Without them speaking their language, I could determine that they're from this particular country.*

As Nesrin walks the streets of Auburn meeting up with *other people from other countries,* she can't help but notice how different each person is. What seems also to be emerging for her is the experience of relationship: *I think it is very nice way of experiencing, finding out how they live and what they do and how they look.* There is pragmatism to relationships. It is only in terms of a relationship that it is possible for newness to become integrated, to be identified and articulated into the potential of the already existing traditions of the society. While the idea of tradition does evoke a past, it also involves the future.

Reflection and change

Through this research I hope to explore what Jack Mezirow (1995) describes as the learning that usually results from a "disorienting dilemma." Such learning can be triggered by a life crisis or major life transition. Such a crisis or transition may well be immigration in its various and different manifestations: as a refugee, as a family reunion or as part of the annual intake of newcomers. For those who "host" the newcomer the disorientation can be just as perplexing.

AUDREY: *If they come over here to have a better life, they should try and live like Australians do.*

Tradition has to be considered as partially determined, as something "already given" – the English language of Australian society, for example. So, the future, for people like Nesrin, is emerging here in the present as she establishes relationships between Turkish and Australian English speakers, as well as with the wider community. In turn, there can be no claim to a purity of tradition from the wider community because the newness that is emerging is no longer commensurate with a particular version of tradition. Our multicultural society cannot generalise itself in terms of a singular tradition.
"Clearly languages and cultures vary in their vulnerability to social and environmental change" (Rhydwen, 1995, p. 8). Homi Bhabha speaks about living the locality of culture:

> This locality is more around temporality than about historicity: a form of living that is more complex than "community," more symbolic than "society"; more connotative than "country"; less patriotic than *patrie*; more rhetorical than the "reason of state"; more mythological than ideology; less homogenous than hegemony; less centred than "the citizen"; more collective than "the subject"; more psychic than civility; more hybrid in the articulation of cultural differences and identification.
> *(1990, p. 292)*

THANH: *Mostly I like Australian food best, chips, hamburgers, pizzas yeah, that sort of thing. I don't think of myself as Australian, not really. I think myself mostly Vietnamese. But there is a side of me that is Australian.*

The social ecology of language and culture in a city environment re-works the present as a site of great variety. The social intensity that accompanies this variety will, no doubt, create social disturbance. At the same time this intensity can generate an internal cohesion and force which provides direction for the society. The consequence of this direction is fundamentally important. To acquire the logic of a multicultural society, difference has to be understood as a complexity, an ontology of difference, which sanctions a plurality of interpretations. Bateson (2019) alludes to this plurality as if complexity is added to complexity, generating mutual learning: the transcontextuality that she calls "between-ness."

The challenge is to partially accept the pragmatism of shared traditions: what is to be incorporated, what will be retained, what might be left out? The intention here is not to turn against, or reject, tradition by denying its presence – such a process would be a futile gesture, only possible through some systematic forgetting. As one can see in contemporary multicultural Australia, any insistence on forgetting has rebounded, bringing with it a reactivation of memory. The recovery of Australian indigenous stories, language and struggle is testament to this. Active memories are essential in the definition and redefinition of how this society goes about holding a plurality of identities.

NESRIN: *Myself, I think I am more a Turk than being an Australian. I could say I value my customs, my background. As well, I do care about Australia as much as I care about Turkey. It's become two nationalities for me. I can stick up for Turkey whenever I need to, and I can also stick up for Australia whenever I need to. I can speak about both countries and defend both countries in every way I need to.*

Nesrin seems to be saying there are soft boundaries to her sense of identity. She can establish for herself an identity that is both Australian and Turkish. The threat of cultural difference, Homi Bhabha (1990) says, is no longer a problem of "other" people; "it has become a question of the otherness of the people-as-one" (p. 301). Nesrin's experience points to the capacity of memory and cultural allegiance to tie a person strongly to another place. She accepts for herself a marginal integration. The learning that this multicultural experiment expects of us is to become comfortable with our shared "otherness."

Local language communities are "communities in the present": *my mother tongue is always in my ears*. But they can lay down no destiny "… Ideally, they (the language communities) 'assimilate' everyone, but holds no one, and it affects all individuals in their inner most being" (Balibar, 1995, p.166).

As Nesrin passes through the education system, she finds herself assimilated but not held to a particular language destiny. She has gone back and forth to Turkey a number of times, living in what Bhabha (1990) calls "the structure of ambivalence that constitutes modern society" (p. 298). In Auburn, the customs and language of particular ethnic communities survive because people live them. There are language-specific newspapers and cultural events, ethno-specific food shops, doctors, dentists and IT shops. The streetscape and shop signage identifies for everyone the great diversity of active languages, and, of course, the library houses large collections of foreign language books, journals and newspapers.

Strong, local ethnic community languages have been able to provide for people the significance of a personal sense of identity. This can motivate people's decisions about making an effort to maintain language and helps them construct a safe, familiar community.

KOKSAL: *Yes, the Turkish language is very important. After all, everything that I learn is from the language, or the background I have. So it's very important. I'm still teaching my daughter the basic Turkish bits and pieces. She start going to Turkish school on weekend now, Friday nights, yeah. Yes, you can say the culture, the religious part and the language makes the Turkish. It's got to be combined together, it's part of a family. It's part of society, or part of backgrounds, that's what it is really. It is sort of combined in little bits and pieces and builds everything up.*

Distortions

There are many issues to consider when thinking about specific language and cultural maintenance: the attitude of parents, job opportunities and street life. Among the people living in Auburn, there are a number for whom speaking their language is inextricably linked into their identity, but there are others for whom it is not. It always is an effort to transmit a language just because it might be useful at some point. People mostly do it because it is important to define who they are, what they stand for, what they believe in. These are ideological reasons and they can create problems because of the intrinsic difficulty of leading an ideologically determined way of life. The current debates and dilemmas that are sometimes referred to as the "Culture Wars" – and the common criticism of certain ethnic communities, often labelled as "ghettos" – can stem from what seems like ideological intransience. On the other hand, it may well simply be an expression of the inherent difficulty in learning a new language.

Some people set out to deliberately lose their language and culture because they just get "fed up" with being made to "speak." If maintaining an ethnic identity, for example, is not particularly important to someone, she is unlikely to choose to speak the language. A personal identity is never fixed, and the fluidity is ever, increasingly, a matter of choice.

MEJ: *When I speak with my Cantonese friends, like, I speak English to them. We speak some Cantonese, like to associate with Chinese, but I can't speak Cantonese fully. I prefer English. I just speak Cantonese to buy things in Chinese shops.*

For some people it is very important to move in and out of these continually created language cultures, occupying several identities within the same person. Cultural identity is a complex phenomenon and it raises the question of "who has the power?" Comprehending one's place or locality, and one's culture, is a relative matter today. English is not the prerogative of native English speakers. Therefore, whose place and culture is it? How does an English speaker address the power of the English language, as well as its powerlessness, standing before a Turkish or Arabic shopkeeper? The complexity of this social ecology is that the shared ground is unruly and asymmetrical. For many migrant communities, language and culture are largely lost or radically altered in the first and second generations. This makes the policy, and even the dream, of cultural pluralism unlikely. But any hope of a seamless assimilation is also unreasonable. The mainstream is continually being hybridised, changing itself in the process. This type of cultural change is transformative. Most non-English speaking migrants become, to some extent, bilingual.

CUNG: *There is a problem for us. We don't want our children to forget totally their own language, their mother tongue. But young people are changing. Because they study in Australian schools and of course they study the way of living here. But when we talk together to each other we sometimes mention something about our country, but we have a problem. We can see that a gap between the old people and the young people, especially the young who were born in Australia or came to Australia at an early age. Because the difference in culture creates some kind of confusion, misunderstanding between the two generations.*

The complex hybrid connections and interdependencies between individuals, the particular language and cultural customs they know, the larger community, and other less tangible aspects of the environment necessary to the survival of culture, can be thought of as an ecosystem, an intricate system of diverse social and emotional factors that interact. This interaction is not static but is continually changing.

> The closed horizon that is supposed to enclose a culture is an abstraction. The historical movement of human life consists in the fact that it is never utterly bound to any one viewpoint, and hence can never have a truly closed horizon. The horizon is, rather, something into which we move and that moves with us. Horizons change for a person who is moving.
>
> *(Gadamer in Snodgrass, 1992, p. 88)*

Stephen FitzGerald (1997) is quite enthusiastic about the way he thinks our contemporary society and culture is changing its historical horizon. He says, Australia "is more vibrant than that of almost any other contemporary culture in the Asian region" (p. 70). Immigrants have brought over 100 languages to Australia. This brings great economic and cultural advantages for everyone.

SERKAN: *I think it is important for me to keep using Turkish. But I will try to learn another language, like Chinese, or Hong Kong, or Japanese, yeah. This is the biggest foreign language. Japanese, they go into markets, like, you know, they own all the markets. Toyota, like all the car factories back in Turkey they own most of the car factories here too. So, I think I will learn Japanese.*

"This is, if you like, the Australian Asian, which has little to do with race" (FitzGerald, 1997, p. 70).

GUNGOR: *I can tell you, living is good here compare to the other countries. But it is a new country this one. It needs time to settle down. So now what we are getting is every culture. Every culture, good or bad it doesn't matter they are bringing here and we are right in observing this one. But the time will come, all the bad things will go out probably, good things remain and we will develop some culture here, our new Australian culture. But will take some time this one. Not a few years, maybe hundred years, my opinion.*

Perspective transformation

At a public forum in Sydney in 2019 on the topic "Learning Together in Living Systems," Nora Bateson stressed both the need to be learning in a communal way and as individuals, learning how to "inner direct" our emotions and cognitive skills. Communities learn in a transformative way, not necessarily because of a described pedagogy, but rather, as this chapter suggests, by an engaged conversation between people, often with very different outlooks. Mezirow (2012) argues that transformative learning occurs in "one of four ways: by elaborating existing frames of reference, learning new frames of reference, transforming points of view, or by transforming habits of mind" (p. 84). The language communities of Auburn are changing cultures. They hybridise and function as ecosystems in relationship. Because complex systems involve so many different people, they can do nothing other than transform, with new frames of reference, altering points of view and creating different habits of mind. The people in this research are in cultural relationships that are structured inside a network of contracts, regulations, expectations, dreams, fantasies, memories and desires. So, at any given moment, strategies available to the individual form a kind of imaginary culture-scape: a domain of metaphor, facts and fictions, memories and fantasies.

At the same time, and alongside the much vaunted gains of multiculturalism, there are also the real effects on the lives of people who feel they have been ignored in our hybridised, pluralist and diverse society. We see this in the so-called populist revolts that are being exploited and harnessed in recent political debates. A nostalgia for a lost past, sometimes thought of as homesickness, isn't just a recent migrant's predicament. The common experience for Australians – apart from, of course, that of the Indigenous peoples of this land – is that of leaving behind, in another place, some qualities of identity and origin. We are a people who have "become Australians." The common condition of our being is that of being near or far from somewhere else: Italy, the United Kingdom, Vietnam, Turkey, Lebanon, Poland, Somalia, Fiji, Korea.

NHAN: *For me, Sydney becoming my home, but actually this is a very big thing for me to decide. At first when I'm staying in Sydney I don't feel like Sydney's really my home. But when I left Sydney, for example, I was in Thailand for some time for holiday, I just realised that Sydney's my home, because Thailand did not relate anything to me. I did not miss much about my old country, like Vietnam, but I miss a lot about Sydney. It is the way we are living here.*

Like Nhan, we too as educators are obliged to engage and learn about ourselves within the awkwardness of our peculiar cultural ecosystem. It is an ever shifting perspective that transforms what it means to become, listen and speak to each other as Australians

References

Balibar, E. (1995). Fictive Ethnicity and Ideal Nation. In J. Hutchinson, & A. D. Smith, (Eds.), *Ethnicity*. Oxford University Press.

Bateson, G. (1972). *Steps to an ecology of mind: Collected essays in anthropology, psychiatry, evolution, and epistemology*. University of Chicago Press.

Bateson, G. (1979). *Mind and nature, a necessary unity*. Bantam Books.

Bateson, N. (2016). *Small arcs of larger circles: Framing through other patterns*. Triarchy Press.

Bateson, N. (2019). *Learning together in living systems* [Presentation]. University of Technology Sydney, Sydney, NSW.

Bhabha, H. (1990). *Nation and narration*. Routledge.

Carter, P (1992). *Living in a new country*. Faber and Faber.

Elias, D. (1997). It's time to change our minds: An introduction to transformative learning. *ReVision*, 20(1), 2–6.

FitzGerald, S. (1997). *Is Australia an Asian country?*Allen and Unwin.

Indyk, I. (2008). *Inner nebulae*. Giramondo Publishing.

Makasa, D., (2008). Beyond my African consciousness. In I. Indyk (Ed.), *Auburn Letters*. Auburn Poets and Writers.

Mezirow J. (1991). *Transformative dimensions of adult learning*. Jossey-Bass.

Mezirow, J. (1995). Transformation theory of adult learning. In M. R. Welton (Ed.), *In defense of the lifeworld*. SUNY Press.

Mezirow, J. (2012). Learning to think like an adult: Core concept of transformation theory. In E. W. Taylor & P. Cranton (Eds.), *The handbook of transformative learning: Theory, research and practice* (pp. 73–115). John Wiley & Sons.

Rhydwen, M. (1995, June 30–July 2). *Why did we give up our Language?* [Paper presentation]. Symposium on Language Loss and Public Policy, Linguistic Institute of the Linguistic Society of America, University of New Mexico, Albuquerque, NM.

Snodgrass, A. (1992). Asian studies and the fusion of horizons. *Asian Studies Review*, 15(3), 81–95.

Stewart, B. (1999). *It ain't where you're from, it's where you're at* [Doctoral dissertation, Western Sydney University]. Western Sydney University Thesis Collection. https://researchdirect.westernsydney.edu.au/islandora/object/uws%3A250

14

HAVE YOU EVER FOUND A GAWURAA?

Christy Hartlage and Jo Clancy

Have you ever found a Gawuraa?

Have you ever found a Gawuraa and wondered whose it was?
Did you keep it; did you leave it?
Did you put it in your pocket?
Did you think about who owned it?
Did you feel it? Did you drop it?
Have you ever found a Gawuraa and wondered whose it was?

Jo

The intention of this piece of writing is to hold space where we, as women, moving, can discuss the gestures that we hold, consciously and unconsciously. As a First Nations woman I am connected and accountable to country and to my community as I dance and create in this magnificent and ancient place. I am grateful to live, create and be embraced in Darug and Gundungurra country.

Christy

We come together through movement and a love of dance. Jo creates a safe space for women to dance together. We have become familiar very quickly. Jo said that one of her mentors told her that she couldn't just dance with women. "But I can and I have." Jo holds a dance space that communicates her experience and her care. We have babies and toddlers, and partners who are leaving, or we may be thinking about leaving ourselves. In a safe place with women, we can leave the mess and dirty nappies and arguments and chaos and the demands of work and just move. When women dance together they heal. They create a place of strength, literally and figuratively. Our spirits and our bodies become physically stronger. In that physicality they can hold the tenderness they need to care for their children, their parents and their communities, and their gestures can hold and heal the wounds of women who have cared in these ways before them. I value the opportunity to move and dance with other women. I value the connections and friendships I have made in getting to know Jo and her community. As a recent resident of the Blue Mountains, I value the way that Jo's dance practice connects me to the stories of this place, to the movements of the birds and the trees here and to the stories that animate the land that I walk.

Both of us

Creativity and movement are essential for all of our lives; we would like to hold a space for this discussion that encompasses tenderness and strength, history and possibility. This kind of space is important for us to understand our place in the world. The value of creative movement rests in a place that enables us to test our bodies, to move ourselves through emotional, habitual and inherited gestures and to find new ways of being. We are connected to women whose movements are as mundane and essential as baking bread and washing clothes. We are connected to them through our physical articulation of joy and tenderness, and of fear, anger and loss, and as we move together we hold that. We are interested in the intelligence of the body, in the way that movement in itself can recreate and regenerate. It is an articulation without words – a different intelligence. We can understand ourselves in relationship to country differently when we understand how to move in response to the other living beings in that space. This movement enables a recognition of the gesture and a curiosity that moves us into that even more. Jo writes about the waradahs (waratahs) that grow over the sand circle in her backyard that "remind us about the many creation stories we have to teach us that from hardship, tragedy and loss, great and beautiful things grow." Our bodies know this, and as we create and recreate the gestures of our grandmothers, we express the meaning of their lives, and the lives of the waradah, the dragonfly and bowerbird, the oak and the bluejay, as they are held in our bodies. Alice Walker (1992) writes, in a dedication to her ancestors, "Rest. In Peace in me the meaning of our lives is still unfolding. Rest." These relationships are played out in our movement, and in the way that we tell the old stories and the stories we tell to make meaning of the joys and challenges of the world we live in now. We cannot respond with the care and creativity required of us unless we can feel the meaning and the resonance of the gestures of our grandmothers, of all of those living beings who came before us that we hold in our bodies. The meaning of their lives is still unfolding. Have you ever found a feather and wondered whose it was?

Jo

I am a mother, a daughter, a sister and an aunty to many. I am a dancer and a choreographer and my life centres around my dance practice. I've been an independent dance artist for 25 years, but have been dancing my whole life. I make dance with and for my community, and as a legacy for the continuation of New South Wales (NSW) Aboriginal dance and storytelling. My practice draws upon my culture, my connection with people and my connection to place. Our interactions with country are continually changing as a result of colonisation and constant new technologies, but the stories held in country can connect us to the dance and movement articulated by our ancestors. I draw inspiration from people and places I feel open-hearted with, and I work with children, young people, dancers and artists who are in turn open-hearted with me.

My dance collective *Wagana* means "to dance" in my paternal grandmother's Wiradjuri language, and it has a strong female youth focus. The work that I make with Wagana resonates and connects with children and young people. My young dancers identify with their Aboriginal heritage through their families and their community, but often their strongest connection to culture comes from dancing with Wagana. These young women and girls are learning everyday who they are, where they connect and their roles in community. They dance, they weave and they sing in language. This helps them join their spirit with their mind and their bodies. Our young people need to feel safe and strong in culture.

I built a sand circle in my back yard five years ago. Here we dance and sing, weave together, eat together, learn and share lessons. Waradahs grow over the sand circle and remind us about the many creation stories we have that teach us about life, its hardships, tragedies and losses, and about the great and beautiful things that can grow from this. Scores of birds fly over, come to rest and drink water from the small pools that form on the dance ground. Girawi (cockatoo) drop pinecones and gather when we dance their dance, a family of gugubarras (kookaburras) visit in the early mornings, and wibigang (magpies) sing and leave us feathers. A dhala-rug (wattle bird) came to nest last spring when I was making a new dance about her for our small girls, and two mulbirrangs (rosellas) have visited my yard, gifting us their wings and their beautiful red and blue feathers. Mugii (owl) has also spent time watching over the dance ground.

There's not a week goes by that Wagana aren't dancing somewhere. Our workshop and performance programme has a repertoire of over 20 dances We dance in schools and pre-schools, at national and international festivals, at corporate and community events. Our dance in education show is called *Gawuraa*, which means feathers. In this show we use song, dance and puppets to take the audience on a journey through country, sharing stories about many of our birds and offering lessons for looking after country, caring for self and caring for others.

> Have you ever found a Gawuraa and wondered whose it was?
> Did you keep it; did you leave it?
> Did you put it in your pocket?
> Did you think about who owned it?
> Did you feel it? Did you drop it?
> Have you ever found a Gawuraa and wondered whose it was?

These Wagana dances come from an embodied and spiritual knowing that I have. It's a different process to my contemporary dance making, which often responds to a piece of music, an idea, a theme decided by someone else or a task-based process with others. Wagana dances, and more recently Wagana songs, come to me in dreams, they come through deep listening when I walk and when I move, and they come on long drives traveling Wiradjuri country. They come from a connection to my grandmothers, my great grandmothers and undoubtedly all of the great women from my family and clan who came before me.

Christy

I am a mother, a partner, a daughter, a sister, an aunty and a good friend. I was born in the United States of America, in San Angelo, Texas. I grew up on a small farm in Western Pennsylvania. When I was growing up we spent a lot of time playing outside, amongst trees and with animals. I love dancing and reading and cooking. I did my Master's degree in the Philosophy of Education at the University of Auckland, and I lived in New Zealand for 12 important years of my life. I danced as a way to balance my headspace while I was studying. I learned to know myself as an independent adult. I fell in love with rocks, and felt a deep connection to that place. I learned the importance of building a caring community around myself. Now I live in the Blue Mountains in Australia with my partner and our two beautiful boys. I have come to know myself as a mother and a partner. I have learned how to build community, and in building community I find myself drawn again into a deeper connection with place. I have lived more than half my life in a place different to where I was born.

The problem with moving and belonging to many places is that there is always a longing for somewhere else, a homesickness that can't really be cured. When I am in America I love the connections with my family, and I would love my children to be able to have a closer connection to their cousins, aunties and uncles. The pace is faster, always moving. I would love to have more time with my brothers and my sister. When I find a quiet place, I relish memories of the scent of maple trees. I remember and I long for the smell of snow and cold that is sweet and makes a certain pain in my sinuses. The energy feels endless and striving. There are possibilities there, realised and unrealised. When I think of New Zealand I long for friends, and the quiet, the pace, comfort and nourishment I feel there. In New Zealand, the land is young, and I feel the green, juicy, youthful energy of Earth that is still forming. I love that. It is vibrant and open. I live in Australia now, and here I value my family and the community around us. I love the feeling of history and life and grounded community that I experience here. Here, with my family and with my friends' families, we care for each other in a way that is timeless, complex and beautiful. In Australia I feel the beautiful, ancestral rocks holding me like a grandmother. I feel the experience and the wisdom of Earth. It is solid here, and strong. Jo said to me it is a privilege to go to bed and wake up in this ancient place, this place that holds the stories of her ancestors. I feel that, and I am drawn to the stories and the healing that I am privileged to breathe in every day.

The first time I went to New Zealand I had been travelling and studying for six months and I was exhausted. I went for a walk on Takapuna Beach. I climbed up onto a massive rock lapped by the waves. As I lay on my back and watched the clouds around Rangitoto, I felt very deeply that I was being cradled by the rock. It was as if the rock was holding my body and soothing it. I felt that I could rest and that it would be fine. I went back to that rock most days in the two months I was there, and each time I felt a connection and an acceptance that I had never felt before. I felt that I had come home. I felt that I could slow down, that the pace of my body fitted with the pace of the rock and the sea. I felt that I belonged in a way that I had not felt before. I fell in love with the rock. In Maketu, I fell in love with the rocks again. I loved watching the way the landscape shifted around them as the tides and the seasons changed. I danced and moved on the beach with the rocks. I walked and cried and swam with the rocks. I felt that I had found my place and my people. I tried to learn Maori and worked to protect the land and the beach and the sea so that this beautiful community could be protected.

Jo

My strongest memory of both my grandmothers is of their smell, their touch and their gestures. My maternal grandmother (my Nan) smelt of Cedel hairspray. Her hands were smooth and tender and her face shone when I entered the room. She was very thin and I remember the wind could blow her over. I cherish her cups of tea. My paternal grandmother (Granma Clancy) had a soft brown face and her dark eyes were framed by thick eyebrows. She smelt like tea and Sao biscuits. She sat low in her hips and was physically strong. I look like Granma Clancy.

I'm a lean woman but often appear big. I have strong legs and also sit low in my hips. I like to move on the floor. I always dance in bare feet and have cracked heels and hard soles. I have a blessed life, with an abundance of good food, clean water and a peaceful home. I dance every day, I sing songs in my grandmother's traditional language and, unlike her, I am not fearful of this. I dance for her and for all the women in my family who were fearful in the wake of colonialism.

Christy

I can feel in my body the gestures that you carry when you teach dance. And I feel he strength of the statement "I dance for her and for all the women in my family who were fearful in the face of colonialism." Your grandmother held and moved with that fear in her body, but she must have been so strong and resilient, and you move with that heritage. It is very clear. Dancing with you I can also feel the joy and pain that I carry from my grandmothers, my own story. I remember my great grandmother, who died when I was 15, telling me about burying her son. She didn't say it, but I felt from her body a howl, a folding in on herself from grief. We are connected to our grandmothers through the gestures of anger and fear and joy and healing tenderness. As we move together we can hold that. What we carry in our bodies comes from our grandmothers, both our human ancestors and the more than human (Abram, 1996) – the rocks and trees and living beings around us that are part of us and that our bodies respond to so strongly. We talk about family resemblance, but there is more than that. It is easy to see the gestures of our ancestors when we perform the tasks that they did to care for us, when we move to nurture our own families and friends. There is more than that though. We also carry the gestures of their pain, their fears and joys. We carry the movements of their bodies over the land on which they walked. And our ancestors are also more than human. We carry in us the sinuous bend of a branch, the solidity of rock, the grace of a falling leaf. These gestures are also carried in our bodies and we repeat them in our responses to place, Earth, country. In our relationship with Earth, that is so primary, essential. This, to me, is the place of movement, of dance. This is the creative response that lives within us, a communication with ancestors and with place that is part of us, always. Movement is a creative intelligence that is essential for understanding how we situate ourselves in relationship to place. The way that your feet caress the ground as you dance, the turn of one's head, the turn of phrase that we use unconsciously that breathes the rhythm of the breeze through gum trees. All of this is part of us.

Jo

Connection to country is strong, particularly through my skin, and the dance that responds to my connection to certain places feels embodied – like it has been danced before. When I was making a dance about galin-balgan-balgang (dragonfly) I visited a large colony of dragonflies. I went many times and sat by the water. My cheeks and fingertips felt cold. I stood quietly in the water they flit across, by the reeds. I cried often. Galin-balgan-balgang visited me in my dreams and in my waking. This dance came slowly and gently and I later understood the importance of this gradual revealing. Darug song woman and dear friend Jacinta Tobin created a song, which she sang in a cycle of three, and it felt like together we remembered something that had been done before.

Both of us

Winhanga-duri-nya is Wiradjuri for reflection, deep listening and meditation. The practice of winhanga-duri-nya can influence our response to crisis. We are facing increasing global challenges with climate change, and we need to develop social practices that encourage a cultural shift to more sustainable living and effective climate action.

Winhanga-duri-nya is the process of bringing our attention to the moment, to be present and non-judgemental, and this in turn can help us be more compassionate to each other, the environment and ourselves. Winhanga-duri-nya can offer us motivation to contribute to solutions.

Before colonisation we had many strategies that responded to our country's weather variations, which we can learn from today. Our traditional ways of governance, which included living communally and sharing resources, meant that we never took more than we needed.

Just as we hold in our bodies connections to our grandmothers, the gestures we hold from our ancestors and from the places we have lived and loved, we also hold grief, fear and anxiety about the future, about climate change and loss of species, the heat and dust of bushfires. This is also part of us. Dance is a practice of listening and responding with openness and tenderness to our own emotional responses. It is creative rather than destructive. Our movements and gestures allow us to find places of hope and reflection in order to make meaning of the world we find ourselves in. Dancing together I feel the strength of the stories that you give us, Jo, and I appreciate the understanding and the connection that we all make together. I think this movement towards a deeper relationship and deeper listening can enable a more vulnerable and honest response to the challenges we all face.

Christy

David Abram (1996), in his book *The Spell of the Sensuous*, writes about the reciprocity of our relationships with nature. Our experience of touching a tree is also the tree's experience of being touched; when we swim in the sea the water moves to accommodate the shape of our bodies; as we breathe in the scent of spring blossoms our breath becomes the breeze that carries the blossoms to the ground. I think that dance can bring us into a stronger awareness of these relationships. Dance requires us to attune to the environment as we move through it. We can look to the curves of a feather as it flutters to the ground, and to the movements of animals for inspiration.

Jo's Bowerbird dance invites us into the bower, to become birds and to play together: to welcome, to gather our treasures, to chase. We watch the bowerbird with curiosity and love, getting to know his movements, recognising his feathers. I feel the muscles in my thighs as we bounce and hop through our bower. I am interested in collecting blue treasures to make our bower more beautiful. I am aware of the bowerbird watching us from the nearby bush, listening to our song. I move towards and away from the other dancers inhabiting the space. These gestures bring me into relationship with the bowerbird that steals the blue pegs from my washing line, and with the other dancers.

Jo

I have another song and dance that tells a story of our women cutting grasses together, and then washing and grinding the seeds to bake bread in campfires. The song came to me when I was sitting by a billabong on Darug Country, and the dance followed very quickly. My body knew how to dance this ritual of gathering, preparing and making bread. It is a commonly held belief that the Egyptians were the first bakers of bread, and many Australians (myself included) were taught this in high school during Ancient History classes. "Wigay" is the Wiradjuri word for bread. It's a word that has always been a part of our ancient language – because we were bakers! 36,000-year-old grindstones have been discovered in New South Wales, used by Aboriginal people to turn seeds into flours for baking. That's well ahead of other civilisations, including the Egyptians, who began making bread only around 17,000 BC (Pascoe, 2014).

Christy

I have my great grandmother's pastry recipe. Every time I make a pie I measure: flour, salt, butter, water – the same ingredients that my great grandmother used, and that my aunties and my cousins use. I measure and mix the same quantities, in the same order. The end of November is Thanksgiving in America and my father's family, the Hartlages, stay together for the long weekend to cook and eat and play together. I make pie on that weekend in Australia, and I am very aware that my aunties and cousins in Ohio are mixing and rolling the dough at the same time that I am, preparing for the Thanksgiving meal. Performing the same gestures. I know what the pastry should feel like, and I move through the process without thinking too much about what I am doing. Instead I am thinking about the preparations for the meal, the jokes and card games, the football on the TV in the background. I think about the importance of that yearly gathering to acknowledge family relationships and tensions. I think about what it means to my children to know that there are people who they are connected to on the other side of the world. I imagine my aunties doing the same as they form the dough into a ball, let it rest and roll it into pie plates. Our family has been doing this for generations. The movement occupies our bodies, my hands remember the texture of this dough and compare this to the dough that I first mixed with my great grandmother when I was six years old. In peace, in me, the meaning of our lives is still unfolding. I long for my family and the ritual of making pies gives me a tangible connection to them.

Jo

Wagana is a space for Aboriginal women and girls to come together in ways Aboriginal women and girls have come together for a long time. Mothers, daughters, sisters, aunties, cousins and kin all dancing together, all learning together, all sharing together, just as our grandmothers did. Wagana women and girls connect to culture through their families and through community. But for some, Wagana is their strongest connection to culture and it links them to cultural practices and traditions. Sixteen-year-old Wagana dancer Anastasia Vickers says "I feel so grounded in my culture when I dance with Wagana. It helps build my confidence and my cultural identity."

Christy

I so admire the way that you write about the connection to country and the depth of connection that resonates with your words and your movement. You asked me what it is like to live far away from the place I grew up in. I feel that I am just being introduced to this place. Your dance practice clearly holds your country, and it is a beautiful introduction. I feel now that I am part of this community; I feel that I belong – it is hard work sometimes. There is a longing in my body, in my bones, for a place where I will not have to explain myself. But I also recognise that when I love a rock, I love the solidness and the tenderness of it, holding and grounded, and I can feel myself breathe and become present. Maybe for me, belonging is about being present and open to the place where I am.

It's about relationships again and connectedness. As we move together we create new stories in response to each other and to the personal and ecological challenges that we face. These new gestures allow us to move with our own vulnerability and find hope and joy. Dancing does that.

Jo

Dance has taken me to many places in Australia and overseas and I enjoy the connections it enables me to have with people and communities. But I miss home. I was raised and still live on Darug and Gundungurra country in the Blue Mountains and I feel a strong sense of place here. I breathe in deeply when I return and feel a gentle hum run through my body. I belong here.

References

Abram, D. (1997). *The spell of the sensuous: perception and language in the more than human world.* Random House.

Pascoe, B. (2014). *Dark emu.* Magbala Books.

Walker, A. (1992). *Her blue body everything we know: Earthling Poems 1965–90 Complete.* The Women's Press.

15

SUSTAINABILITY WORK

An urgent need for a new profession

Werner Sattmann-Frese and Stuart B. Hill

Introduction

We are arguing here for the need to create the new profession of ***sustainability worker***, one dedicated to emotionally and psychosocially enabling individuals in personal sessions and psychoeducational group meetings to live sustainable lives with small ecological footprints. We also envisage that the required qualifications of ***sustainability workers*** would enable them to work as advocates for communities and as educators in the development of governmental and corporate social and environmental responsibility.

The theories of change[1] and consciousness development underpinning this proposal are informed by social ecological thinking, which posits that our ability to live is dependent on our ability to connect issues relating to mental and physical health with ecosystem health (Hill, 2011; Sattmann-Frese & Hill, 2009). This way of making sense of living is still only embraced tentatively in research and practice.[2] We consider that ecological problems, such as extreme weather events, not only exacerbate mental health problems, but are also the results of trauma-based self-esteem issues, and the widespread compulsion to compensate a lack of self-esteem through the addictive and unsustainable consumption of goods and services.

In this chapter we:

- argue that the world is getting worse at many interrelated levels;
- argue that the current approaches have not met our expectations for change, because of their lack of attention to the psychological underpinnings of ecological problems;
- describe some key aspects of unsustainable living;
- suggest that the creation of deep change will require large numbers of people to engage in effective and comprehensive personal healing journeys;

- describe the key aims and tasks of this proposed profession of ***sustainability worker;***
- describe the skills and training required to become a professional ***sustainability worker*** and;
- examine possible public motivation for using what these professionals have to offer.

The world is getting worse

In their book *We've Had a Hundred Years of Psychotherapy and the World's Getting Worse*, Hillman and Ventura (1993) note that the work of psychotherapists has not been able to significantly contribute to reducing the world's problems. Also disturbing is the realisation that the efforts of environmental thinkers, deep and social ecologists, and environmental and sustainability educators,[3] and even transformative educators,[4] have also not been able to significantly reduce the ecological decline. And, despite the excellent contributions of certain environmental ethicists, eco-theologians and ecopsychologists[5] to help people deepen their understanding of the links between psyche, soul and the world, they also have had less than the expected effects on the current dominance of unsustainable lifestyles. However, an increasing number of ecologically aware psychotherapists are now engaged in enabling individuals to live more sustainably.[6]

Despite the efforts of probably millions of people worldwide, ecological progress has been slow, and has mainly involved technological advances and curative (back-end) initiatives rather than the more needed wellbeing-enabling and problem-prevention (front-end) whole-system redesign. In terms of the six-step model of eco-consciousness development proposed in our 2009 book *Learning for Sustainable Living*, public discourse remains stuck in the early stages of consciousness development. Most programmes still rely on managerial and technological interventions, and on educational efforts that do not take account of the depth psychological "trauma medication" underpinnings of unsustainable and harmful perceptions and behaviours.

From a social ecology perspective, it is not only concerning that the notion of consciousness development is still largely neglected in discourses about the future, but also that there has been a decline in many key aspects of sustainable living, including global inequality, right-wing fascism and deterioration in mental health (even in countries, such as Australia (CSIRO, 2018), with a relatively well-organised health system). Key concerns include the following:

In many countries, the output of greenhouse gases is continuing to rise – as is the denial by politicians that this is a man-made problem – and that we are now living in the Anthropocene, with humans having significant impacts on the Earth's geological and ecological systems (Hamilton, 2017).

Powerful corporations and politicians continue to act as obstructers to, and postponers of, the much needed positive ecological and psychosocial change. Climate denial think tanks and misinformation campaigns have been secretly funded by such corporations as Koch Industries, who since 1997 have given over $100,000,000 to 84 groups in the USA who deny climate change science.[7]

Mental health in Western countries continues to decline. According to the World Health Organisation (2017), "depression will be the number one health concern in both the developed and developing nations by 2030."[8] According to the Australian Government Report on the Health of Children and Adolescents (Lawrence, et al., 2015), "[a]lmost one in seven (13.9%) 4 to 17 year-olds were assessed as having mental disorders in the previous 12 months."

Twenge, Zhang and Im (2004) have found that there is an overall sense of loss of control of our fate, with a shift from the importance of internal (intrinsic) goals towards an external focus and orientation to life (extrinsic goals). And there has been a dramatic increase in addictions among 45 to 54 year-olds, with associated dramatic increases in mortality from substance abuse, destructive behaviour and suicide.[9]

Addressing the causes of such psychosocial (and ecological) decline will require a radical redesign of systems at all levels, to avoid rather than just ameliorate problems. The extent of what is required, organisationally and psychosocially, has been illustrated by one of us in relation to the design and management of sustainable food systems (Hill, 2014).

In our view, ***sustainability workers*** would work particularly with individuals and groups for whom the satisfaction of basic material needs is not a main issue, but who (mostly subconsciously) are using the satisfaction of unsustainable wants, not needs, to cope with past and present emotional distress, and to bolster their diminished sense of self.[10] An important focus of ***sustainability work*** will be to enable these people to become aware of this, and to lead satisfying lives without "needing" to resort to compensatory consumption and distraction to regulate their emotions and bolster their self-esteem. We envisage that ***sustainability workers*** will also be employed by communities as psychosocially aware sustainability educators and advocates. They will also have the skills to systemically work with families, work groups and political stakeholders, helping them to transform both their "consumption and production behaviours" in integrated ways, thus avoiding the current distractive debates about what must change first: consumption or production.

We believe that if we do not manage to stop ecological and social decline in the next few years, we will have eventually destroyed the world. While we may have satisfied our immediate wants, we will have neglected our deeper need to do the work of developing an ecologically sustainable personal sense of self, and of caring for the planet, one another and its non-human inhabitants.

Unsustainable living

Although the number of individuals engaging in lifestyles of health and sustainability (LOHAS)[11] is increasing, most are still engaged in unsustainable consumption patterns that harm their personal health and wellbeing, impacting the natural environment through the unsustainable use of resources, including through excessive processing, packaging, transportation and storage. Many of these individuals may also smoke, consume excessive amounts of alcohol and lead inactive lifestyles.[12] With increasing age, these are also the individuals most likely to suffer from a diverse range of minor and major ailments.

The above-described patterns of behaviour are key drivers of our increasing environmental and social decline.[13] This "affluenza" is the painful, contagious, socially transmitted condition of overload, debt, anxiety and waste, resulting from the addictive pursuit of more (de Graaf, et al., 2014; Hamilton & Dennis, 2005). Such unsustainable consumption may include unnecessarily expensive houses, cars, electronic gadgets and clothes, which are often psychologically branded to convey to the buyer a sense of "being someone."[14] Such "retail therapy" is really "addictive retail acting out," because the external satisfaction of symbolic wants through compensatory consumption can only ever provide short-term relief from feeling emotionally low and unimportant in a society that undermines people's sense of self (by, for example, increasing inequity and then promoting consumption as the fix).

At the same time, "self-regulation" through the use of pharmaceutical products, rather than emotional living skills, has become the new way of keeping increasing levels of anxiety and depression under a thin veneer of control. Even in France, where there is a higher degree of work–life balance than in the USA and Australia, nearly one in four people are taking pharmaceutical products – such as tranquillisers, antidepressants and antipsychotics – to regulate their emotions and get through the day.[15]

Skills and training

What kind of skills would a ***sustainability worker*** have to acquire to be able to help individuals to effectively address the deep underpinnings of this harmful and unsustainable condition? Considering the emotional, psychosomatic and psychosocial breadth and depth of peoples' struggles with sustainable living, briefly outlined here, we believe that such workers would need to attain the following competencies:

- Understand key aspects of social and environmental deterioration, including the effects of climate change, overconsumption and the various forms of emotional, social and ecological exploitation;
- Understand the emotional and psychodynamic underpinnings of unsustainable perceptions and behaviours in general, and compensatory consumption in particular;
- Be familiar with and able to effectively use – in their roles as counsellors, advocates and educators – supportive knowledge and practice frameworks, including those developed in social ecology, deep ecology, ecofeminism and ecopsychology;
- Have advanced counselling and self-regulation skills to be able to assist individuals in healing emotional wounds caused by past and present insults to the self and from emotional deprivation;

- Have the facilitation skills to work with groups of people in ecologically effective ways in societies that will in future have to be spending increasingly more money and effort on climate change mitigation and associated environmental problems;
- Have skills to enable people to enhance their experience of body–mind and "person–planet unity";
- Have a working knowledge of relevant movements and initiatives, such as "The Slow Moment"[16] and "Lifestyles of Health and Sustainability" (LOHAS)[17];
- Be able to help people suffering from illnesses to make sense of their physical and emotional suffering, e.g. psychosomatics, critical and positive psychiatry, holistic understandings of addictions;
- Have a holistic understanding of the socio-political factors involved in enabling genuinely progressive cultural change and the facilitation skills needed to enable this;
- Have the design and redesign skills needed to focus on enabling change at the front-end of the sustainability and lifestyle challenges we face, rather than just at the back-end (where the usual problem-solving initiatives only address symptoms instead of causes).

What, then, would the work of a ***sustainability worker*** look like? A typical professional would probably work at local community centres in the service of environmentally progressive local councils, providing structured sustainable living seminars and workshops.[18] Participants would likely be individuals interested in enhancing their abilities to lead more conscious and satisfying lives with a small ecological footprint – individuals who would welcome opportunities to connect with like-minded others and be open to sharing resources and significantly changing their ways of living.

Unlike the current sustainability education approaches that assume that it will suffice to provide information on unsustainable and sustainable behaviours, ***sustainability work*** considers unsustainable perceptions and behaviours in the context of people's very personal psychosocial and psychosomatic struggles with sustainable living. This enabling of personal growth, and an increase in inner harmony and sense of connectedness, is likely to make unsustainable behaviours *redundant*. In our view, this is an even deeper change than increases in "willingness to change" to implement sustainability behaviours in personal and family life, as described by Léger (2011), and to engage in what Chris Riedy (2016) calls interior transformation.

Sustainability workers would also be able to work with those living with addictions, who would benefit from developing a personalised and ecologically aware approach to growing out of their addictions and associated ecologically unsustainable mindsets and behaviours. Carers of mental health consumers and people working in the disability sector could also benefit from professional development seminars that link the strains that come with their profession with their overall ability to lead more sustainable lives.

Sustainability workers may also choose to work in private practice to enable individuals to make sense of the distresses experienced at work, home and in built environments, and from a wide range of ecological, psychosomatic and lifestyle perspectives. Such counselling would be supported by body–mind and mindfulness modalities capable of deepening an embodied sense of self. This work may be viewed as a specialised (more personalised and "ecologised") branch of social work and provided by graduates with a Master's degree in Social Work with a Major in ***Sustainability Work***, or with a specialisation in counselling and psychotherapy studies.

Motivation

Motivation to seek the support of ***sustainability workers*** is likely to be driven by a need to consciously enhance one's mental and physical wellbeing and, at the same time, reduce one's negative impact on the natural environment. Participants may also want to survive social, emotional and economic hardship, and to avoid sliding into difficult-to-manage states, such as homelessness and social disintegration. Presently an increasing number of financially well-resourced people (including CEOs of corporations, medical doctors, psychologists and psychiatrists) are seeking the support of counsellors and psychotherapists who are using such approaches as psychodynamic psychotherapy, dialectical behaviour therapy, mindfulness-based cognitive behaviour therapy (CBT) and somatic (body-oriented) psychotherapy.[19] Also, an increasing number of celebrities have been seeking therapeutic support to deal with the idiosyncrasies of their profession and status,[20] and many among the super-rich – who appear to be increasingly struggling with shame, isolation and other symptoms of the so-called "wealth fatigue syndrome" – are also seeking support.[21] Whether or not the healing experienced by such people can reduce the exploitation of ordinary citizens remains controversial.[22] The social-ecology-informed view on ecological and social change promoted here certainly suggests that it will need consciousness change at a grassroots level, not just philanthropic schemes and "wealth therapy" to achieve ecological and psychosocial sustainability.

The following three groups of potentially receptive individuals may be recognised:

Those struggling with satisfying their basic needs who hope that access to goods and services will enable them to lead satisfying lives. These are likely to remain the typical clients of social workers.

Those who are relatively financially secure but still hope that the better job, the bigger house and car, and some other trappings will enable them to feel more content and satisfied and develop a stronger sense of self. They may need to be motivated through community campaigns to seek the support of ***sustainability workers***.

Those who have already learnt that the things money can buy do not make them happy or solve their problems but may have significant harmful effects on their health and wellbeing. Many may already be working with professionals to enable them to address the past and present causes of their emotional struggles.

From our years of teaching social ecology at Western Sydney University, and ecologically aware counselling and psychotherapy at the Jansen Newman Institute in Sydney, we are confident that the above-mentioned individuals could benefit from the more ecologically oriented approach to counselling, psychotherapy and group work that ***sustainability work*** would emphasise.

The following three fields of theory and practice (education for sustainable development, green social work and ecotherapy) are complementary to the proposed work of ***social workers***.

Education for sustainable development (ESD) and environmental education

The UN Decade of Education for Sustainable Development charged educators with developing and implementing educational programmes for "securing sustainable life changes, aspirations, and futures for young people" (Pavlova, 2013). Key sustainable development issues such as biodiversity, climate change, poverty reduction and sustainable consumption were commonly included in school sustainability curricula. The emphasis has been on teaching, mainly school children, about the importance of environmentally sustainable perceptions and behaviours, assuming that this will nurture more biocentric understandings and lifestyles, with reduced ecological footprints (Gough, 2014; Littledyke et al., 2009). We question the assumption that awareness, on its own, leads to action (Firth & Smith, 2017). Subconscious "adaptive voices/selves" are always likely to undermine and sabotage this connection, so actions tend to be symbolic, publicly acceptable, downsized or postponed rather than being what is really needed.

So, predictably the effectiveness of this "informing" approach has been limited, as acknowledged in the UNESCO publication "A decade of progress on sustainable development: Reflections from the UNESCO chairs programme" (Michelsen & Wells, 2017).

> Despite a decade of ESD momentum, and despite it being well over twenty years since the first Earth Summit in Rio de Janeiro, we find the state of humanity and the planet in continued decline. The urgency of finding a response is greater than ever.
>
> *(p. 18)*

Although the UNESCO text mentions the importance of self-awareness, global awareness, environmental and health literacy (p. 23), discussion of the development of an eco-self-consciousness is conspicuously absent, and the authors conclude that the "future requires a rethinking of each of our institutions, in context. It involves thinking about institutional policies in more harmonious and coordinated ways" (Michelsen & Wells, 2017, p. 79), and the focus has been on programmes in schools, with the task of "saving the planet" being left to the next generation (Saylan & Blumstein, 2011).

Although some adult learning programmes are being developed by organisations such as **Adult Learning Australia**, [23] adult learning for sustainable development has been significantly left behind (Orlovic Lovren & Popovic, 2017).

Ecological social work and ecopsychology

Although there are individuals in both of these areas with concerns that overlap ours, until positions for ***sustainability workers*** are established – with access to appropriate resources and supports – we consider that it will not be possible to design and implement effective programmes to enable the cultural transformation required for achieving genuine ecological sustainability.

Green social workers (Dominelli, 2012; Kapro, 2016; Ramsay & Boddy, 2017)[24] can be called on to provide support in responding to environmental crises and also for their expertise in designing programmes that integrate responses to other social issues, including those relating to wellbeing, equity and social justice.

Ecopsychologists (Clinebell, 1996; Robinson, 2009; Smith, 2015)[25] can similarly contribute, by sharing their understanding of our psychosocial evolution, ways to enable the development of deep connections (and reconnections) with nature, and the formation of an ecological self.

Conclusions

We consider that ***stainability workers*** could contribute to Ted Trainer's (2010) version of a "Conserver Society." Despite the limited success from more than 100 years of psychotherapy and many decades of sustainability education, we hold out hope that a more ecologically, psychosomatically and psychodynamically aware approach to working for sustainable living, in tandem with existing ecotherapy and sustainability education practices, can make a significant difference within the windows of opportunity that we still have available to us to avoid personal, social and planetary collapse.

We believe that the integrated approach we are advocating to support personal mental and physical health, and ecosystem health, and its provision by appropriately trained professionals, is what is needed now: by enabling the global development of ecological consciousness. Without these changes we will be destined to spend more and more of our resources on mitigating climate change and social destruction – and a declining human population – outcomes that the contributions in this book are aiming to help avoid.

Selected list of organisations and programmes of relevance to the development of *sustainability workers*

Adult Learning Australia: https://ala.asn.au/education-for-sustainable-development/

California Institute of Integral Studies: https://www.ciis.edu/academics/course-descriptions/toward-an-integral-ecological-consciousness

Ecopsychology: https://www.teachgreenpsych.com/ecopsychology/

Institute for Social Ecology – Online Seminar: Social Transformation Beyond Pragmatism or Utopia: http://social-ecology.org/wp/2018/09/online-seminar-social-transformation-beyond-pragmatism-or-utopia/

International Community for Ecopsychology: http://www.ecopsychology.org/

Keele University – Master of Science in Environmental Sustainability and Green Technology: https://www.keele.ac.uk/greenkeele/keelehub/mscenvironmentalsustainabilityandgreentechnology/

Lewis & Clark Graduate School of Education & Counselling – Ecopsychology Certificate: https://graduate.lclark.edu/departments/counseling_psychology/ecopsychology/

Naropa University: Master of Arts in Ecopsychology: https://www.naropa.edu/academics/masters/ecopsychology/index.php

New Ecopsychology: http://www.new-ecopsychology.org/

Pacifica Graduate Institute – Certificate in Ecopsychology: https://retreat.pacifica.edu/certificate-in-ecopsychology/

Prescott College – BA in Ecopsychology: https://www.prescott.edu/academics/concentrations/ecopsychology

Prescott College: Post-Masters Certificate in Ecotherapy: https://online.prescott.edu/online-certificate-programs/ecopsychology

Project NatureConnect.com: Applied Ecopsychology and Ecotherapy: http://www.ecopsych.com/pnccom/

Western Sydney University: Master of Education (Social Ecology) https://www.westernsydney.edu.au/future/study/courses/postgraduate/master-of-education-social-ecology.html

Whitman College – Environmental Humanities: https://www.whitman.edu/academics/departments-and-programs/environmental-studies/the-environmental-studies-major/environmental-humanities

Wilderness Reflections - Professional Ecotherapy Certification Program: https://www.wildernessreflections.com/clinical-ecotherapy-certification-program/

Notes

1 See also Learning for Sustainability: http://learningforsustainability.net/theory-of-change/ and Andrea A. Anderson: The Community Builder's Approach to Theory of Change: http://www.theoryofchange.org/pdf/TOC_fac_guide.pdf

2 Two examples here are Sandifer, Sutton-Grier & Ward (2015) and Lindgren & Elmqvist (2017).

3 Environmental educators such as Berry (1988/2015); Blewitt & Tilbury (2013); Capra & Luisi (2014); Conn & Conn (2008); Diesendorf (2009); Drengson & Devall (2015); Fien (2010); Gough (2014); Hill (2012); Jickling & Sterling (2017); Kahn (2009); Mulligan (2017); Norgaard (1995); Orr (2016); Palmer (1998); Robottom & Stevenson (2012); Schumacher (1978); Sessions (1995); and Tilbury, Adams & Keogh (2005).

4 Transformative educators such as Boyd & Myers (1988); Cranton (2016); Daloz (1999); Dirkx (2009); Hill, et al. (2004); Kasl & Yorks (2002); O'Sullivan & Taylor (2004); Bardaglio & Putnam (2009); and Wright, et al. (2013).

5 Authors in this group include Bucke (1901/2011); Buzzell & Chalquist (2009); Fisher (2013); Glendinning (1995); Kasser & Kanner (2003); Koger & DuNann Winter (2010); Lasch (1991); Macy (2014); Plumwood (1991); and Roszak (2003/1978; 2001).
6 Examples of ecopsychologists are provided by the International Community for Ecopsychology: http://www.ecopsychology.org/
7 Koch Industries: secretly funding the Climate Denial Machine (2015): http://www.greenpeace.org/usa/global-warming/climate-deniers/koch-industries
8 Facts and figures about mental health (2014): https://www.blackdoginstitute.org.au/docs/default-source/factsheets/facts_figures.pdf?sfvrsn=8
9 See for example Kuerbis et al. (2014).
10 See also Brad Tuttle's 2010 article entitled "Psych Study: When You're Bummed, You're More Likely to Buy": http://business.time.com/2010/05/07/study-low-self-esteem-makes-you-more-likely-to-buy-luxury-goods/
11 See for example http://www.lohas.com.au/what-lohas
12 See for example Paul Zollinger-Read (2013): How technology and inactive lifestyles are changing our children: https://www.theguardian.com/sustainable-business/technology-inactive-lifestyle-changing-children
13 See also: Miles (2010).
14 See for example Sivanathan & Petitt (2010).
15 See for example businessculture.org (2018). Work-life balance: https://businessculture.org/western-europe/business-culture-in-france/work-life-balance-in-france/
16 See for example Ann Handley (2018): Rethinking ASAP: the magic of going slow at just the right moments: https://annhandley.com/rethinking-asap/
17 http://www.lohas.com.au/;https://en.wikipedia.org/wiki/LOHAS; see also Cultural Creatives: http://culturalcreatives.org/; & Ray & Anderson (2001)
18 PowerPoint presentations on such initiatives can be viewed at https://www.slideshare.net/WernerSF
19 See for example Smita Navare (2008): Counseling at work place: A proactive human resource initiative: https://www.ncbi.nlm.nih.gov/pmc/articles/PMC2796765/; Amy Morin (2015): Why highly successful people seek therapy: https://www.forbes.com/sites/amymorin/2015/05/28/why-highly-successful-people-seek-therapy/#324e99de3145
20 Iva Anthony (2014): Celebs who have gone to therapy: https://madamenoire.com/452098/celebs-who-have-gone-to-therapy/; Lindsay Holmes (2017): 12 celebrities get real on the power of going to therapy: https://www.huffingtonpost.com.au/entry/celebrity-therapy-quotes_us_5981e1cce4b0353fbb33f42b
21 Jana Kasperkevic (2015): Wealth therapy tackles woes of the rich: "it's really isolating to have lots of money": https://www.theguardian.com/us-news/2015/oct/17/wealth-therapy-tackles-woes-of-the-rich-its-really-isolating-to-have-lots-of-money; and Nick Duerden (2015): A new breed of "wealth therapists" is here to help the super-rich: https://www.independent.co.uk/life-style/health-and-families/features/a-new-breed-of-wealth-therapists-is-here-to-help-the-super-rich-a6701781.html
22 See for example Helaine Olen (2015): Why you should never, ever, ever take "wealth therapy" seriously: https://slate.com/business/2015/10/wealth-therapy-is-bogus-and-so-are-its-roots-in-financial-therapy.html; and Kate Aronoff (2015): Wealth therapy is an insult to us all: meet the 1 percenters finding solace in wealth redistribution: https://www.salon.com/2015/10/24/wealth_therapy_meet_the_1_percenters_finding_solace_in_wealth_distribution_partner/
23 https://ala.asn.au/
24 See also her 2013 PowerPoint Presentation: Green social work and environmental justice in an environmentally degraded, unjust world: https://www.ulapland.fi/loader.aspx?id=738c09c1-fa9b-4475-af63-a506967870e1
25 What is ecotherapy? http://www.ecotherapyheals.com/; see also Buzzell & Chalquist (2009).

References

Australian Government Department of Environment and Heritage (2005). *Educating for a sustainable future: A national environmental education statement for Australian schools.* Curriculum Corporation. http://www.seedengr.com/sustainable-future.pdf

Bardaglio, P., & Putman A. (2009). *Boldly sustainable: Hope and opportunity for higher education in the age of climate change.* National Association of College and University Business Officers.

Berry, T. (2015). *The dream of the Earth.* Counterpoint LLC. (Original work published 1988).

Blewitt, J., & Tilbury, D. (2013). *Searching for resilience in sustainable development: Learning journeys in conservation.* Routledge.

Boyd, R. D., & Myers, G. J. (1988). Transformative education. *International Journal of Lifelong Education,* 7(4), 261–284.

Bucke, R. M. (2011). *Cosmic consciousness: A study in the evolution of the human mind.* White Crow Books. (Original work published 1901).

Buzzell, L., & Chalquist, C. (Eds.) (2009). *Ecotherapy: Healing with nature in mind.* Sierra Club Books.

Capra, F., & Luisi, P. L. (2014). *The systems view of life: A unifying vision.* Cambridge University Press.

Clinebell, H. (1996). *Ecotherapy: Healing ourselves, healing the Earth.* Routledge.

Conn, L., & Conn, S. (2008). As inside, so outside: Restoration of inner and outer landscapes. In R. L. France (Ed.), *Healing natures, repairing relationships: New voices on restoring ecological spaces and consciousness.* Libri Publishing.

Cranton, P. (2016). Transformative learning: a narrative. In M. J. Spector, B. B. Lockee, & M. D. Childress (Eds.), *Learning, design, and technology: An international compendium of theory, research, practice, and policy.* Springer.

CSIRO (2018). *Future of health: Shifting Australia's focus from illness treatment to health and wellbeing management.* CSIRO. https://www.csiro.au/en/Showcase/futureofhealth.

Daloz, L. (1999). *Mentor: Guiding the journey of adult learners.* Jossey-Bass.

de Graaf, J., Wann, D., & Naylor, T. H. (2014). *Affluenza: How overconsumption is killing us – and how to fight back* (3rd ed.). Berrett-Koehler Publishers.

Diesendorf, M. (2009). *Climate action: A campaign manual for greenhouse solutions.* UNSW Press.

Dirkx, J. M. (Ed.). (2009). *Adult learning and the emotional self: New directions for adult and continuing education.* Jossey-Bass.

Dominelli, L. (2012). *Green social work.* Polity Press.

Drengson, A., & Devall, B. (Eds.). (2015). *The ecology of wisdom: Writings by Arne Naess.* Counterpoint LLC.

Fien, J. (2010). *Work, learning, and sustainable development.* Springer.

Firth, R., & Smith, M. (Eds.). (2017). *Education for sustainable development: What was achieved in the DESD* [the UN Decade of Education for Sustainable Development]. Routledge.

Fisher, A. (2013). *Radical ecopsychology: Psychology in the service of life* (2nd ed.). SUNY Press.

Glendinning, C. (1995). Technology, trauma, and the wild. In T. Roszak, M. E. Gomes, & A. D. Kanner (Eds.), *Ecopsychology* (pp. 41–54). Sierra Club Books.

Gough, S. (2014). *Education, nature and society.* Routledge.

Hamilton, C. (2017). *Defiant earth.* Allen & Unwin.

Hamilton, C., & Dennis, R. (2005). *Affluenza: When too much is never enough.* Allen & Unwin.

Hill, S. B. (2011). Social ecology: An Australian perspective. In D. Wright, C. Camden-Pratt, & S. B. Hill (Eds.), *Social ecology: Applying ecological understanding to our lives and our planet* (pp. 17–30). Hawthorn.

Hill, S. B. (2012). Education to change the world: Learning for realising one's personal, social & ecological potential. *21st Century Learning.* https://www.researchgate.net/publication/

302906192_Education_to_Change_the_World_Learning_for_Realising_One's_Personal_Social_Ecological_Potential

Hill, S. B. (2014). Considerations for enabling the ecological redesign of organic and conventional agriculture: A social ecology and psychological perspective. In S. Bellon & S. Penvern (Eds.), *Organic farming: Prototype for sustainable agricultures* (pp. 401–422). Springer.

Hill, S. B., Wilson, S., & Watson, K. (2004). Learning ecology: A new approach to learning and transforming ecological consciousness: experiences from social ecology in Australia. In E. V. O'Sullivan & M. Taylor (Eds.), *Learning toward an ecological consciousness: Selected transformative practices* (pp. 47–64). Palgrave Macmillan.

Hillman, J., & Ventura, M. (1993). *We've had a hundred years of psychotherapy and the world's getting worse*. Harper.

Jickling, R., & Sterling, S. (Eds.). (2017). *Post-sustainability and environmental education: Remaking education for the future*. Springer/Palgrave Macmillan.

Kahn, R. (2009). *Critical pedagogy, ecoliteracy, & planetary crisis: An ecopedagogy movement*. Peter Lang.

Kapro, K. (2016). What is green social work? *Social Work Helper*. https://www.socialworkhelper.com/2016/10/13/green-social-work/

Kasl, E., & Yorks, L. (2002). Toward a theory and practice for whole-person learning: Reconceptualizing experience and the role of affect. *Adult Education Quarterly*, 52(3), 176–192.

Kasser, T., & Kanner, A. D. (Eds.). (2003). *Psychology and consumer culture: The struggle for a good life in a materialistic world*. American Psychological Association.

Koger, S. M., & Du Nann Winter, D. (2010). *The psychology of environmental problems: Psychology for sustainability* (3rd ed.). Psychology Press.

Kuerbis, A., Sacco, P., Blazer, D., & Moore, A. A. (2014). Substance abuse among older adults. *Clinical Geriatric Medicine*, 30(3), 629–654.

Lasch, C. (1991). *The culture of narcissism: American life in an age of diminishing expectations* (Rev. ed.). W. W. Norton & Company.

Lawrence, D., Johnson, S., Hafekost, J., Boterhoven de Haan, K., Sawyer, M., Ainley, J., & Zubrick, S. R. (2015). *The mental health of children and adolescents: Report on the second Australian child and adolescent survey of mental health and wellbeing*. Commonwealth of Australia. http://www.health.gov.au/internet/main/publishing.nsf/content/9DA8CA21306FE6EDCA257E2700016945/$File/child2.pdf

Léger, M. T. (2011). A grounded theory perspective on eco-sustainable change in families. *Ecopsychology*, 3(4), 237–247.

Lindgren, E., & Elmqvist, T. (2017). Ecosystem services and human health. *Oxford Research Encyclopedia of Environmental Science*. doi:10.1093/acrefore/9780199389414.013.86

Littledyke, M., Taylor, N., & Eames, C. (2009). *Education for sustainability in the primary curriculum: A guide for teachers*. Palgrave Macmillan.

Macy, J. (2014). *Coming back to life: The updated guide to the work that reconnects*. New Society Publishers.

Michelsen, G., & Wells, P. J. (Eds.). (2017). *A decade of progress on sustainable development: Reflections from the UNESCO Chairs Programme*. UNESCO.

Miles, S. (2010). *Spaces for consumption*. Sage.

Mulligan, M. (2017). *An introduction to sustainability: Environmental, social and personal perspectives* (2nd ed.). Taylor & Francis/Routledge.

Navare, S. (2008). Counseling at workplace: A proactive human resource initiative. *Indian Journal of Occupational and Environmental Medicine*, 12(1), 1–2.

Norgaard, R. B. (1995). *Development betrayed: The end of progress and a coevolutionary revisioning of the future*. Routledge.

O'Sullivan, E., & Taylor, M. (Eds.). (2004). *Learning toward an ecological consciousness: Selected transformative practices*. Palgrave Macmillan.
Orr, D. (2016). *Dangerous years: Climate change, the long emergency, and the way forward*. Yale University Press.
Orlovic Lovren, V., & Popovic, K. (2018). Lifelong learning for sustainable development — is adult education left behind? In W. Leal Filho, M. Mifsud, & P. Pace (Eds.), *Handbook of lifelong learning for sustainable development*. (pp. 1–17). Springer.
Palmer, J. A. (1998). *Environmental education in the 21st century: Theory, practice, progress and promise*. Routledge.
Pavlova, M. (2013). Towards using transformative education as a benchmark for clarifying differences and similarities between environmental education and education for sustainable development. *Environmental Education Research*, 19(5), 656–672.
Plumwood, V. (1991). Nature, self, and gender: Feminism, environmental philosophy and the critique of rationalism. *Hypatia – A Journal of Feminist Philosophy*, 6(1), 3–27.
Ramsay, S., & Boddy, J. (2017). Environmental social work: A concept analysis. *British Journal of Social Work*, 47(1), 68–86.
Ray, P. H., & Anderson, S. R. (2001). *The cultural creatives: How 50 million people are changing the world*. Random House.
Riedy, C. (2016). Interior transformation on the pathway to a viable future. *Journal of Future Studies*, 20(3), 35–54.
Robinson, L. (2009). Psychotherapy as if the world matters. In L. Buzzell, & C. Chalquist (Eds.), *Ecotherapy: Healing with nature in mind* (pp. 24–29). Sierra Club Books.
Robottom, I., & Stevenson, R. B. (2012). Analyses of environmental education discourses and policies. In R. B. Stevenson, B. Robert, M. Brody, J. Dillon, & A. E. J. Wals (Eds.), *International handbook of research on environmental education* (pp.123–125). Routledge.
Roszak, T. (2001). *The voice of the earth*. Phanes Press.
Roszak, T. (2003). *Person/planet: The creative disintegration of industrial society*. Anchor Press/ Doubleday (Original work published 1978).
Sattmann-Frese, W. J., & Hill, S. B. (2009). *Learning for sustainable living: Psychology of ecological transformation.*. Lulu.com.
Saylan, C., & Blumstein, T. (2011). *The failure of environmental education*. University of California Press.
Schumacher, E. F. (1978). *A guide for the perplexed*. HarperCollins/Perrenial.
Sessions, G. (1995). *Deep ecology for the 21st century*. Shambhala.
Sivanathan, N., & Petitt, N. C. (2010). Protecting the self through consumption: Status goods as affirmational commodities. *Journal of Experimental Social Psychology*, 46(3), 564–570.
Smith, J. (2015). *Making sense of ecotherapy*. Mind.
Tilbury, D., Adams, K., & Keogh, A. (2005). *A national review of environmental education and its contribution to sustainability in Australia: Further and higher education*. Australian Government Department of the Environment, Water, Heritage and the Arts and the Australian Research Institute in Education for Sustainability (ARIES). www.aries.mq.edu.au/projects/national_review/index.php#Vol2
Trainer, T. (2010). *The transition to a sustainable and just world*. Envirobook.
Twenge, J. M., Zhang, L., & Im, C. (2004). It's beyond my control: A cross-temporal meta-analysis of increasing externality in locus of control, 1960–2002. *Personality and Social Psychology Review*, 8, 308–319.
World Health Organisation (2017). *Mental health atlas 2017 Member state profile*. https://www.who.int/mental_health/evidence/atlas/profiles-2017/AUS.pdf?ua=1
Wright, D., Camden-Pratt, C., & Hill, S. B. (Eds.). (2013). *Social ecology: Applying ecological understanding to our lives and our planet*. Hawthorn Press.

LIST OF CONTRIBUTORS

Sean Blenkinsop is Professor, Faculty of Education, Simon Fraser University, Vancouver, Canada. He grew up in the boreal forests of northern Canada. With a 30+ year background in outdoor, environmental and experiential education, his interest in wild pedagogies comes quite naturally. Now, as Professor of Education, he has been involved in starting three nature-based, place-based, eco-schools (all in the public system) and has written extensively about these experiences and the philosophical underpinnings of eco-education writ large.

Jo Clancy is a mother, a daughter, a sister and an aunty to many. She is a First Nations choreographer, dancer, teacher and mentor. Jo is a descendant of the Wiradjuri people of Western NSW and was raised and still lives on Darug and Gundungurra country in the Blue Mountains with her family.

Jo commenced her full-time dance training at NAISDA Dance College in 1990 and then went on to Western Sydney University where she became the first Aboriginal person to gain a Bachelor of Arts in Dance in NSW. Jo was Head of Dance at NAISDA from 2005–2007 and is currently the Cultural Dance Programme Coordinator at the College. Jo founded her dance collective *Wagana* in 2007.

Over the past twenty-five years Jo has developed many contemporary Aboriginal dance works and education projects for festivals and events throughout Australia and overseas, including *Bangalang* for the Commonwealth Youth Dance Festival in Glasgow, *Sum of our Ancestors* for Dance and the Child in Copenhagen, Denmark and most recently *Wirawi Bulbwul* for Dance and the Child in Adelaide, Australia.

Ann Dale is Professor in the School of Environment and Sustainability at Royal Roads University and she held her university's first Canada Research Chair in sustainable community development (2004–2014). She is a Trudeau Alumna and a Fellow of the World Academy of Art and Sciences. She is a recipient of the CUFA

Paz Buttedahl Distinguished Career Academic Award (2014), the Canada Council for the Arts Molson Prize for the Social Sciences (2013) and the 2009 Bissett Award for Distinctive Contributions to the Public Sector. Her book *At the Edge: Sustainable Development in the 21st Century* received the 2001 Policy Research Initiative Award for Outstanding Research Contribution to Public Policy. Her research focuses on climate pollution, governance, research curation, sustainable community development, social capital and agency. Ann holds degrees in psychology (1975) and public administration (1994) from Carleton University, and a doctorate in natural resources sciences from McGill University (1999), earning a place on the Dean's Honour List.

Jen Dollin is the Senior Manager, Sustainable Futures in the Office of the Pro Vice Chancellor Learning Futures where she leads a range of sustainability engagement and curriculum renewal initiatives. She is passionate about experiential education that transforms students' relationships with their place and has extensive experience working collaboratively on funded initiatives with university, school, community and government networks in the Greater Western Sydney and the Asia Pacific region. Jen's background in corporate finance triggered her "epistemological shock" and propelled her towards studies and work in environmental economics, environmental policy development and transformative education. In addition to her full-time role, she is currently immersed in the final two years of her PhD in environmental humanities along the Upper Hawkesbury River, her heartland. Her research interests are in the fields of feminist post-humanism, new materialism and multispecies ethnography.

Kate Harris is passionate about authentic leadership, experiential education and meaningful outcomes. Her mission is to create sustainable solutions for our present and future global communities through human capability. Kate is currently involved in leadership development in the Australian Public Service Commission, Department of Prime Minister and Cabinet. She is a past CEO of GECA (Good Environmental Choice Australia), Vice Chair of the Living Future Institute of Australia (LIFA), a non-executive director of ALCAS (Australian Life Cycle Assessment Society) and a non-executive director of Global Ecolabelling Network. Prior to her role at GECA, Kate was the former CEO for the Centre for Sustainability Leadership, a unique and successful programme whose alumni continue to cooperate for great change. A key component of Kate's own learning was through her completion of a Master's degree in social ecology at Western Sydney University. Kate also draws on her diverse professional experience in the performing arts, not-for-profit sector and organisational development.

Christy Hartlage has a Master's degree in Education from the University of Auckland, New Zealand. She has travelled widely, studying sustainable agricultural practices in South America, Thailand, India and New Zealand. She was a witness to the New Zealand inquiry into genetically modified agriculture (having farmed echinacea in the Bay of Plenty region). She is the mother of two boys. In 2010 she established a volunteer home-visiting service (YAWN) that provides practical support for families with babies. This highly respected service brings families into community, with the support of volunteer women and their kindness and care. Christy is an excellent cook and the founder of Madam Lasagna's cooking school for children. She loves teaching people how to cook delicious things and loves to dance.

Roseanna Henare-Solomona is is a Ngati Hine, Ngapuhi woman who has lived in Sydney for 30 years. Her PhD explored the changes to Maori identity and culture after migration to Australia. It also investigated how Maori research methodology and Western knowledge can work together to enable storylines from those who live in and between two worlds: the old and new, Indigenous and non-Indigenous. Roseanna currently lectures at Western Sydney University on global leadership, governance and social enterprise, social ecology and transformative learning, and leadership. Her research interest is about how systems and complexity thinking can inform new ways of seeing and understanding an ever-changing world.

Stuart B. Hill was appointed Foundation Chair of Social Ecology at Western Sydney University in 1996. At WSU he taught qualitative research methodology, social ecology research, transformative learning, leadership and change, and a programme titled "Sustainability, Leadership and Change: A Social Ecology Perspective." He was appointed Emeritus Professor in 2013. His PhD (1969) was a ground-breaking whole-ecosystem study of community and energy relationships.

Prior to 1996 he was at McGill University, in Montreal, where he was responsible for the zoology degree, and where, in 1974, he established Ecological Agriculture Projects, a resource centre for sustainable agriculture (http://www.eap.mcgill.ca/).

Stuart has published over 350 papers and reports. His latest books are *Ecological Pioneers: A Social History of Australian Ecological Thought and Action* (with Martin Mulligan), *Learning for Sustainable Living: Psychology of Ecological Transformation* (with Werner Sattman-Frese) and *Social Ecology: Applying Ecological Understanding to Our Lives and Our Planet* (with David Wright and Catherine Camden-Pratt).

Dale Hunter is an innovator in the area of group facilitation. She is a co-founder and Director of the facilitation company Zenergy, a network of facilitators generating presence and collective intelligence towards a vision of "Whole People Co-operating in a Sustainable World" (www.zenergyglobal.com). Dale is principal co-author of *The Art of Facilitation: The Essentials for Leading Great Meetings and Creating Group Synergy*, *Handling Groups in Action: The Use of Distinctions in Facilitation*, *The Zen of Groups: The Handbook for People Meeting with a Purpose* and *Co-operacy: A New Way of Being at Work*.

Dale is a sustaining member of the International Association of Facilitators (IAF) and was Vice Chair International from 2001 to 2007. She co-championed the development of the IAF Code of Ethics for Group Facilitators and was a recent President and council member of the Peace Foundation of New Zealand/Aotearoa.

Rachael Jacobs is a Lecturer in creative arts education at Western Sydney University and a former secondary teacher (dance, drama and music) and primary arts specialist. Her research interests include assessment in the arts, language acquisition through the arts and embodied learning. Rachael has facilitated arts projects in community settings all over Australia, including in refugee communities, prisons and women's refuges. In 2016 she contributed to the arts education component of the OECD report on Sustainable Development Goals for 2030. She is the former Director of Research for Drama Australia, a community activist, a freelance writer, aerial artist, dancer and choreographer. She was a founding member of the community activism group Teachers for Refugees and runs her own intercultural dance company.

Bob Jickling was appointed Professor of Education at Lakehead University after many years of teaching at Yukon College; he continues his work at Lakehead as Professor Emeritus. He now lives in Canada's Yukon Territory because, from the beginning in 1979, this land has sung to him. He has never abandoned the ideas of wilderness and the wild and has puzzled about this for years. More recently he has found resonant chords in his concerns about education. Along the way, he founded *Canadian Journal of Environmental Education*, co-chaired the 5th World Environmental Education Congress in Montreal, and wrote many scholarly papers. But now he has returned to his Yukon home, to hear it sing, to complete a few projects and to ponder possibilities for "wild pedagogies."

Hilary Leighton is an Associate Professor in the School of Environment and Sustainability at Royal Roads University and a lifelong apprentice to nature and psyche. Hilary draws upon the wisdom of ecopsychology, depth psychology, whole-systems thinking, mythology, embodiment, nature-based and arts-based practices, and poetry. Her research and teaching reflect the ethical dilemma, suffering and loss of our relationships with what is wild and contemplates critically conscious ways of learning as an initiation towards maturation and a more soulful way of belonging to the world.

Hilary is a Registered Clinical Counsellor in British Columbia and a psychotherapist. She is also in private practice. She holds a PhD in interdisciplinary studies (University of Victoria, 2014), a Master of Education in Curriculum and Education (Simon Fraser University, 2004) and certification in Integrative Body Psychotherapy (2007). She was a recipient of the University of Victoria Interdisciplinary Graduate Fellowship Award 2012–14 and founding director of Continuing Studies at Royal Roads (2006–2016).

Cathy McGowan has worked as a regional councillor for the Victorian Farmers Federation and is a former President of Australian Women in Agriculture. She was elected as an independent member to the Australian House of Representatives for the rural Victorian seat of Indi in the 2013 federal election, and served in that position until her retirement before the 2019 federal election. In 2004 she was made an Officer of the Order of Australia "for service to the community through raising awareness of and stimulating debate about issues affecting women in regional, rural and remote areas." She was also a recipient of the Centenary Medal in 2001. She has a Master's in Applied Science in Agricultural and Rural Development from the University of Western Sydney.

Christine Milne, AO is the Global Greens Ambassador. She was appointed following a long career in Green politics in Australia (1989–2015) serving in the national parliament as a senator for Tasmania and leader of the Australian Greens. Passionate about protecting biodiversity and addressing global warming, she drove the introduction of a carbon price and the Clean Energy Package whilst holding the balance of power with the Gillard Labor Government (2010–13). She is patron of the Smart Energy Council of Australia, ambassador for the Invasive Species Council, board member of The Climate Mobilization, advisory board member of Solar Heads of State, and member of Climate Accountability Institute.

Werner Sattmann-Frese currently works as a Senior Lecturer teaching mental health, psychotherapy practice, research, somatic psychotherapy and psychosomatics at the Jansen Newman Institute in Sydney. He studied medicine (1977–1981) and body-oriented psychotherapy (1981–84), and has been working in private practice as a psychotherapist and supervisor for more than 35 years. He holds a Master's of Applied Science and a doctorate in social ecology from Western Sydney University. His research interests focus on the roles of counselling and psychotherapy in a world of ecological decline and on the creation of ecologically aware mental health practices and educational programmes.

Subarna Sivapalan is the director of the Centre for Excellence in Teaching and Learning at Universiti Teknologi PETRONAS, Malaysia. At the national level, Subarna is the co-chair of the National Education for Sustainable Development Work Group. She was also the Chair of the Foundation for Environmental Education EcoCampus Committee from 2018 to 2019. Subarna has been an Adjunct Fellow at Western Sydney University's School of Education since 2016. Her areas of expertise include education for sustainable development, environmental humanities and sustainability communications.

Brendon Stewart joined the School of Social Ecology at Western Sydney University in 1991. He helped develop and implement the first undergraduate degree programme and curriculum for social ecology in Australia. Brendon also engaged in qualitative studies into sense-of-place theory and practice, the latter being successfully carried through by way of place-specific colloquia over a number of years and at different locations across Australia. Theories to do with a sense of place have informed Stewart's primary research, which considers how recently arrived migrants in the Australian society are able to establish a sense of home. Home is central to the study of ecology, and the complexity of adaption to local situations is an on-going social-learning process.

Isak Stoddard is a PhD researcher in natural resources and sustainable development with a particular focus on regional climate and energy transformations. Over the past decade he has worked in various capacities at the student-initiated Centre for Environment and Development Studies (CEMUS) at the University of Uppsala and the Swedish University of Agricultural Sciences. At CEMUS, his worked mainly focused on developing transformative and transdisciplinary approaches to higher education as a response to global environmental and societal challenges. Isak was also involved in the creation of a 10-year research initiative on climate change leadership, centred around a series of visiting professorships at Uppsala University made possible by a donation from Zennström Philanthropies. He has an educational background in engineering physics and energy systems. A dual citizen of Sweden and the United States, he is now based in Uppsala, but enjoys frequent adventures to the Scandinavian mountains for skiing, wandering and climbing.

Ganakumaran Subramaniam is the Head of School and Professor at the School of Education, University of Nottingham, Malaysia Campus. He has been Vice President of AsiaTEFL since 2014. He holds a BEd in TESL from Universiti Putra Malaysia, and read for his Master's and PhD at the University of Nottingham, UK. Prior to joining the University of Nottingham, he served as Acting Director of International Relations at the National University of Malaysia. He was a member of the Malaysian Council for English Language Quality and Standards and is a Fulbright Scholar with attachments in New York, Washington and New Mexico, USA. He is currently the Academic Director of the Bangladesh College Education Development Project (2016–2021). He has published in the areas of teacher education, teaching pedagogies and education policies.

Stephen J. Thorpe is a Senior Lecturer in the Department of IT and Software Engineering, and the Programme Leader of the Master of IT Project Management, at the Auckland University of Technology. He is also the Director of External Relations and Development and leads the student–industry engagement portfolio for the School of Engineering, Computer and Mathematical Sciences. As an academic, educator and team facilitator he brings a unique skill set to the implementation of strategy-aligned initiatives in line with the university's vision. Stephen is also a Director of Zenergy, a network of facilitators generating presence and collective intelligence towards a vision of "Whole People Co-operating in a Sustainable World." His research interests include enhancing online team effectiveness, group facilitation, agile facilitation, eGovernment and IT governance. Recent work includes a co-authored paper entitled "Enhancing the facilitation of online groups in higher education," published in 2019 in *Interactive Learning Environments.*

David Wright is a senior lecturer in Education at Western Sydney University, Australia. He teaches in the overlapping areas of systems theory, social ecology, cross-cultural learning and transformative learning. His research is informed by his interest in embodiment and the body in learning, developed initially through pursuit of a career in writing for performance. In this respect, he is interested in learning processes and the contexts within which learning occurs. These include (our experience of) the changing world environment as a consequence of hyper-capitalism and its contribution to climate change. These and their manifestation in the immediate, the local and the close personal encounters of everyday life are of significance to him. He has published widely on drama and environmental education and is co-editor of *Social Ecology: Applying Ecological Understanding to Our Lives and Our Planet* (with Stuart Hill and Catherine Camden-Pratt).

INDEX

Note: page references in bold indicate tables; italics indicate figures; 'n' indicates chapter notes.

For Product Safety Concerns and Information please contact our EU representative GPSR@taylorandfrancis.com
Taylor & Francis Verlag GmbH, Kaufingerstraße 24, 80331 München, Germany

www.ingramcontent.com/pod-product-compliance
Ingram Content Group UK Ltd.
Pitfield, Milton Keynes, MK11 3LW, UK
UKHW020203200926
475702UK00004B/592
* 9 7 8 0 3 6 7 4 7 1 0 8 8 *